JSL Companion

Applications of the JMP® Scripting Language

Theresa L. Utlaut
Georgia Z. Morgan
Kevin C. Anderson

The correct bibliographic citation for this manual is as follows: Utlaut, Theresa L., Georgia Z. Morgan, and Kevin C. Anderson. 2011. *JSL Companion: Applications of the JMP® Scripting Language.* Cary, NC: SAS Institute Inc.

JSL Companion: Applications of the JMP® Scripting Language

ISBN 978-1-61290-094-0 (electronic book)
ISBN 978-1-60764-880-2

SAS Institute Inc., SAS Campus Drive, Cary, North Carolina 27513-2414

1st printing, December 2011

Contents

About This Book

Purpose

The title that we selected, "JSL Companion: Applications of the JMP Scripting Language," is supposed to be representative of what we want to achieve. Learning is a journey, and a good travel companion can make a journey more engaging and less arduous. A travel companion helps with navigating a route, pointing out not only features of interest, but also how to avoid wrong turns into bad neighborhoods. Our purpose is to provide support for people interested in applying JSL.

Is This Book for You?

There are many great resources for JMP scripting. This book is intended to supplement them and reach a broad spectrum of JMP and JSL users. If you are a novice just learning JSL, if you are somewhat familiar with scripting, or if you want to advance your scripting skills, this book might prove useful. There are many example scripts and applications included with the text that can be used as reference or as a building block for your own applications.

Prerequisites

This book assumes that you have access to JMP and that you are comfortable using JMP interactively, including facility with data tables and the menu structure, and at least some basic familiarity with table manipulation and analysis platforms.

Scope of This Book

This book has characteristics of learning a language by immersion: we use JSL scripts to teach JSL scripting. Topics in each chapter include example scripts. If you can copy portions of these example scripts and use them immediately and productively, even before understanding them completely, then this immersion approach has been successful. Our experience suggests that after JSL novices produce their first useful script, the added confidence is accompanied by an eagerness to read and learn more about JSL.

How This Book Is Organized

The book is divided into 10 chapters with multiple sections in each. The first chapter is introductory and intended for a JSL novice. It covers some of the basics and helps with getting started. Chapters 2 through 5 are what we consider the building blocks of JSL: input and output, working with data tables, script-writing essentials, and JMP data structures. Chapters 6 through 8 are the core for building an application: creating reports, communicating with users, and custom displays. The two final chapters focus on flexible scripting and provide some helpful hints for laying out your scripts, debugging, and improving performance.

Typographical Conventions Used in This Book

Text	is used for text.
Bold	is used for JMP functions, commands, and any JMP interface item used in an example or figure directly associated with the text around it.
Not Bold	is used for variables, filenames, and any JMP interface item in general.

Software Used for This Book

JMP 9 for Windows was used exclusively for the scripts and as the target of all the screen captures in this book. Specifically, the scripts were written and tested in JMP 9.0.3.

How to Use This Book

Navigating

If you are new to scripting, then Chapter 1, "Getting Started with JSL," is the place to begin. If you already have some familiarity with scripting, then that chapter can probably be skipped without missing too much. However, if you have not worked with lists in JSL, then reviewing the section about lists in the first chapter might be beneficial.

The chapters in this book, and even the sections within a chapter, are not really intended to be read sequentially. From personal experience, we realize that few people will read a book like this from cover to cover. For the majority of readers, it will function as a reference book, and it will be used when a new project is started or when an existing script needs debugging. Our advice is to read through salient sections as you need help on a topic, and browse through the others when you can spare some time. JSL is an extensive language, and we tend to learn something new almost every day. Hopefully, you will too.

Running the Scripts

To easily run the sample scripts downloaded from the *JSL Companion* Web site, extract the zipped file of scripts and data to the directory of your choosing.

Log in to JMP, and run the script 0_CreatePathVariables.jsl. The script prompts you to browse to the path where you extracted the scripts and data.

This script creates a JMP path variable named JSL_Companion. The script includes two commands to test that the path has been set correctly and to provide the syntax for using it. Many of our example scripts include Open and Write statements using this path variable.

```
//once this is run you specify this path variable with a leading $
Open("$JSL_Companion/2_ReadData.jsl");
Open("$JSL_Companion/Deli Items.jmp");
```

Ensure the preference JSL Scripts should be run only, not opened… is disabled. Go to **File ▶ Preferences ▶ Windows Specific** and ensure that this preference is unchecked. Most of our scripts assume the default platform preferences are enabled, for example **Show Points**. If your results do not match the results displayed in this book, check your platform preferences.

PC Versus Mac

Because this book was written using JMP on a Windows operating system, when shortcut keys are used or referenced, we assume that you are using JMP for Windows. JMP provides a handy *Quick Reference* guide, available by selecting **Help ▶ Books ▶ Quick Reference**. It includes a long list of shortcut keys for Windows and Macintosh. It is worth a look, regardless of your operating system.

References

Dickens, Charles. 1997. *Our Mutual Friend.* London: Penguin Classics.

Morgan, J. 2010. "Expression Handling Functions: Part I, Unraveling the Expr( ), NameExpr( ), Eval( ), … Conundrum." *JMPer Cable, A Technical Publication for JMP Users*; Issue 26 Winter 2010:15-19.

Murphrey, W., and R. Lucas. 2009. *Jump into JMP Scripting.* Cary, NC: SAS Institute Inc.

NASA Langley Research Center Atmospheric Science Data Center, ISCCP. Percent Cloud Cover Data. Available at http://isccp.giss.nasa.gov/products/products.html.

Rose, David. "A JMP 3-D Graphics Art Collection." *JMP File Exchange.* September 2010. Available at http://support.sas.com/demosdownloads/sysdep_t4.jsp?packageID=000416&jmpflag=Y.

SAS Institute Inc. 2010. *JMP 9 Design of Experiments Guide.* Cary, NC: SAS Institute Inc.

SAS Institute Inc. 2010. *JMP 9 Modeling and Multivariate Methods.* Cary, NC: SAS Institute Inc.

SAS Institute Inc. 2010. *JMP 9 Scripting Guide.* Cary, NC: SAS Institute Inc.

SAS Institute Inc. 2010. *Using JMP 9.* Cary, NC: SAS Institute Inc.

Weiss, E. H. 1985. *How to Write a Usable User Manual,* ISI Press.

Acknowledgments

We wish to acknowledge our family, friends, students, colleagues, employer, publisher, suppliers, and others who have helped make us who we are and, thereby, contributed to this book.

In particular, we are grateful to Stephenie Joyner, our acquisitions editor, who guided us through this process with patience, fortitude, and a sense of humor. Thankfully, she talked us out of living in a van down by the river. This book could not have happened without the help of many others at SAS Press. A very special thank you to Mary Beth Steinbach, Candy Farrell, and Jennifer Dilley for their support and creativity; Shelly Goodin, Aimee Rodriguez, and Stacey Hamilton for their marketing prowess; Amy Wolfe for her perseverance and endurance; and Julie Platt, SAS Press Editor-in-Chief, who provided the opportunity to write this book.

The JMP reviewers improved not only this book with their recommendations, but also our knowledge and understanding. For sharing their knowledge and expert advice, special thanks go to: Mark Bailey, Michael Crotty, Melanie Drake, Rosemary Lucas, Paul Marovich, Tonya Mauldin, and Wendy Murphrey.

Thanks also go to JMP Technical Support for addressing an Avogadro's number of questions.

We offer special thanks to John Sall, first and foremost a statistician, whose foresight, creativity, and dedication established a user-friendly product with a scripting language accessible to the general user.

Theresa wishes to thank Mark and Bandit for their encouragement and patience from start to finish, and the village that it took to raise her: Mom, Dad, Lori, Jim, Karen, Brian, Jane, and all the others.

Georgia wishes to thank Paul, Ted, Alexis, James, and her entire family: your anticipation became my motivation. For Clark, words will never be adequate to express the value of your insight, your words, custom DLLs that enhance JMP features, and encouragement. Finally, many thanks go to her colleagues who pose interesting questions, and to her employer who provides the opportunities to explore them.

Kevin wishes to thank Karen, Charlie, Amanda, Gene, and Ruth for their unwavering support and encouragement. This couldn't have happened without them.

As we undertook this enterprise, one goal to which we committed ourselves was to still be friends when we finished. Though numerous personal and professional challenges arose, we are proud of achieving this goal as well.

1 Getting Started with JSL

Introduction

We don't want anyone to get hurt, so the first chapter warms up the reader with gentle stretching using the JMP Scripting Language (JSL). This chapter demonstrates a portion of the utility of scripting in JMP, using explanations and examples that detail the basics of the language. Then, we introduce more useful and advanced concepts. After a short demonstration showing the vast possibilities of JSL, we cover a few basic concepts, describing some of the windows, effective and efficient script writing from JMP, and preliminary scripting concepts, including punctuation, messages, naming, and lists. This chapter builds a foundation that supports your journey into JSL scripting.

The Power of JMP and JSL

Opportunities to transform data into information come at us every day like a fire hose aimed at a shot glass. We are industrial statisticians, supporting the development and manufacturing fabrication facilities in the technology manufacturing group of a large semiconductor company. We consult with engineers to maximize their returns on investments of time and effort. We teach classes on statistics and experimental design. We try to do something wonderful by finding innovative ways to get valid, actionable information in front of management to better enable its decisions. And, for all of this and more, one of our most useful tools is JMP.

JMP is powerful desktop software that was created by SAS more than 20 years ago "because graphical representations of data reveal context and insight impossible to see in tables of numbers." Its point-and-click interface, capabilities, and style enable analysts without much formal training to make defensible, data-supported recommendations in a short period of time with less effort. JMP is as advertised: visual, interactive, comprehensive, and extensible.

That extensibility comes from JSL. JSL is an interpreted language that can implement the data manipulation and analyses available in JMP in a flexible, concise, consistent, standardized, and schedulable way. Indeed, a talented and motivated scripter can write new analyses, new procedures, or new visualizations that implement methods not available in the point-and-click interface of JMP. The scripter can deploy these methods across an entire enterprise. Through JSL, almost any data manipulation, analysis, or graphic can now be generated, provided enough knowledge, innovation, and perseverance are applied. We are often amazed at the scripts written by our coworkers that demonstrate not only the generation of information from data elegantly, but do so in a manner or sequence that we would not have considered ourselves. Of course, there are some holes in the innate capabilities of any software application, but we believe that the capability of a script is usually only limited by the skill, perseverance, and imagination of the scripter.

If you have some experience with JMP and JSL, you probably already feel this way. Or, you suspect that it's true at the least. We can hear the uninitiated saying, "Wow, the hyperbole meter has hit the peg!" Fair enough. We know the doubters need proof. Hang tight; we provide demonstrations within our JSL applications throughout the rest of this book. But, for right now, let's look at a few samples.

First and foremost, JMP is visual. You might have already peeked at the sample script named Teapot.jsl in the Scene3D folder. Or, you might have looked at the Statue of Liberty script in the JMP 3-D Graphics Art Collection by David Rose from the JMP File Exchange Web site. These two scripts are impressive displays of visual power, even if only for artistic appreciation.

Let's say your manager wants a presentation-ready process-capability report in his inbox every Monday morning. You can take comfort in knowing that this report and the accompanying tabular reports are possible to generate, publish on a Web site, and mail on a scheduled basis using JSL.

Figure 1.1 Capability Report

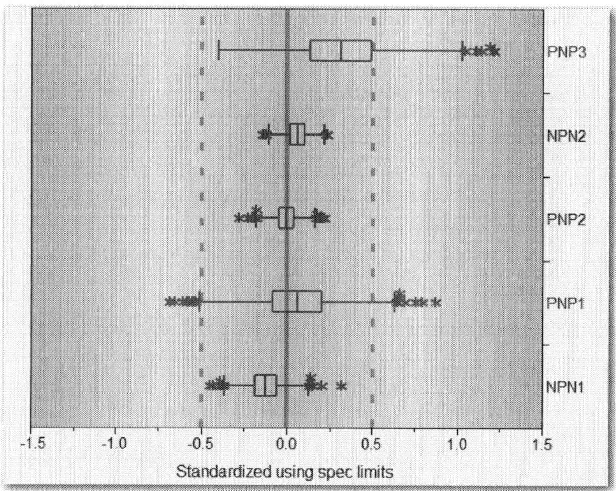

JMP is interactive. Using some basic JSL, dialog boxes can gather salient information from users for deployment in analyses.

Figure 1.2 Custom Dialog Built with JSL

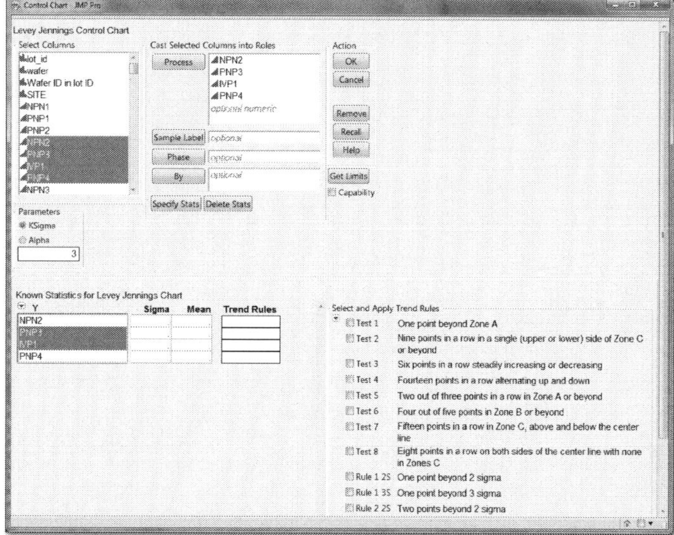

With some JSL, users can interact with graphics through text entry or sliders.

Figure 1.3 Visualizing the Weibull Distribution

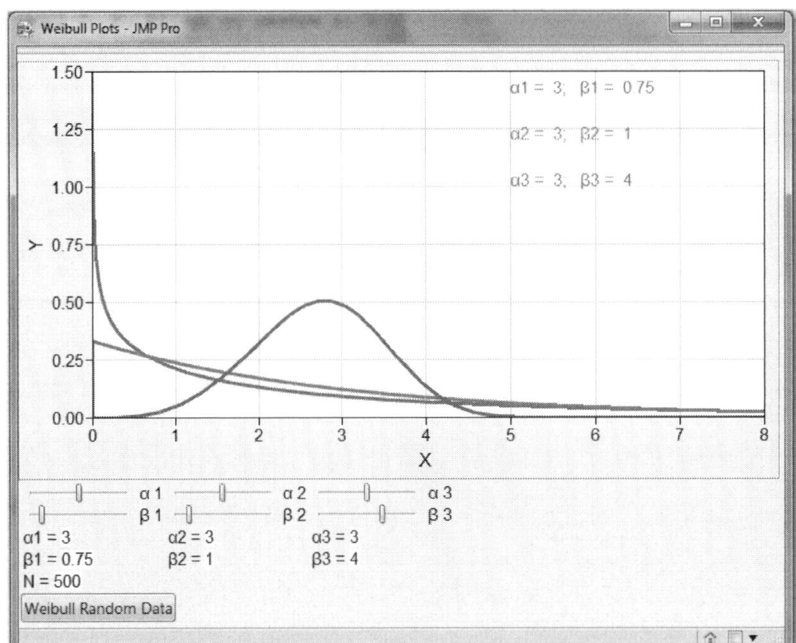

JMP is comprehensive. JSL lets you control most of the innate capabilities of JMP. Even where there are holes in the capabilities of JMP, a wily scripter can use SAS, R, or JSL data, functions, matrix-manipulation abilities, and extensive graphic control to generate and manipulate data sets, perform innovative procedures and analyses, and return the results for display and reporting. Again, your script is limited not by the capabilities of JMP, but only by your skill and imagination.

The Basics

Have you ever had an instructor who started the class with a comment similar to, "You'll have no problem learning this. It's really quite easy"? Isn't that an annoying comment from someone who is an expert? Of course, it seems easy if you already know it. Learning something new can be intimidating and hard. Fortunately, many tasks in JSL are relatively easy. There is no sense in being disingenuous, saying that mastering JSL is simple. It's not. In fact, expert JSL programmers learn how to do something new or optimize a script on a regular basis. With the helpful scripting tools in JMP and a few instructions, useful JSL scripts can be written in a short time. We regularly see students write useful scripts that improve productivity after taking just a four-hour introductory course. We predict that as you write more scripts, you will discover that you have developed a feel for JSL. You will start surprising yourself by knowing commands that you have never used simply because you have an understanding of the structure of the language.

Create and Run a Script

Now that you have warmed up with some stretching, let's do a little exercise. You are going to create a script. It is a simple script, but it will give you a sense of the structure of JSL, and your confidence will build about learning a new language.

In JMP, open a new script window. The Script window is discussed in more detail in the next section of this chapter.

There are several ways to open a new script window in JMP.

- From the menu bar, select **File ► New ► Script**. (See Figure 1.4.)
- From the **Home** toolbar, click the Script icon.
- From the **JMP Starter** window, select **New Script**.
- Hold down the CTRL key, and select T.

Figure 1.4 New Script Window Using the Menu Bar

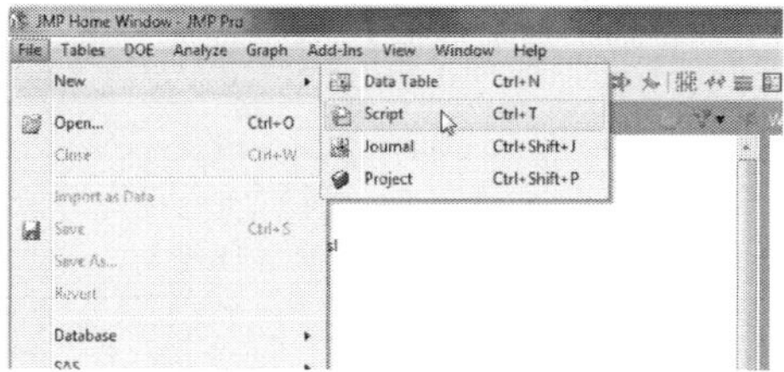

For your first script, type the following code into the script window. Note: All scripts in this section are included in the 1_TheBasics.jsl script.

```
txt = "In teaching others we teach ourselves.";
Show( txt );
```

Now, run your script. There are several ways to run a script in JMP.

- Click the red JMP man icon on the **File_Edit** toolbar.
- Select **Edit ► Run Script**.
- Right-click on the script, and select **Run Script**.

- Hold down the CTRL key, and select R.

You can run portions of a script by highlighting the lines of code to run, and then using one of the previous ways to run just the highlighted code.

After the script is run, it prints the variable name and text in the Log window. If the Log window is not open, select **View ▶ Log**.

```
txt = "In teaching others we teach ourselves.";
```

There are a few important things to note about this simple script:

- The variable txt is assigned the text string using a single equal sign.
- The text string is enclosed within double quotation marks.
- Semicolons follow each line of code and glue them together. Semicolons tell JMP there is more to do (more lines of code). The semicolon in the last line of code is not required, but it does not cause an error if it is included.
- The text enclosed within double quotation marks is magenta in color, and the JSL function **Show( )** is blue. These are the default colors used in the Script window to make the code more readable and easier to debug.
- There are spaces in the **Show( )** function. Extra spaces within or between JMP functions or within JMP words are okay, and they can make the code easier to read. The same is true for tabs, returns, and blank lines.
- The Log window is your friend.

All of these points are covered in more detail throughout the book.

Open, Modify, and Save a Script

In the following example, the JMP Sample Data file CO2.jmp is used. A script opens the data file from the JMP Sample Data file directory. It creates a scatter plot of CO2 versus Year&Month, and then fits a line to the data.

```
CO2_dt = Open( "$SAMPLE_DATA/Time Series/CO2.jmp" );

CO2_dt << Bivariate( Y( :CO2 ),
     X( :Name( "Year&Month" ) ),
     Fit Line()
);
```

Figure 1.5 CO2 versus Year&Month Fit Line

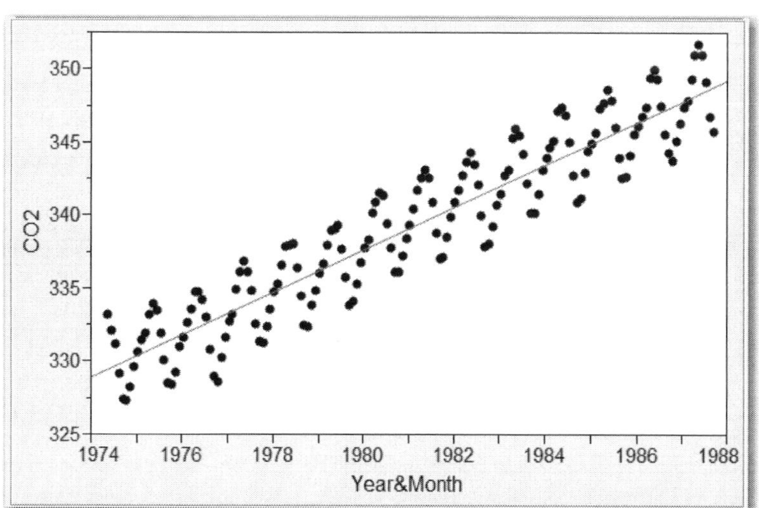

From the scatter plot, you can see that there is a linear structure to the data. Fitting a line does not tell the entire story. There is structure that remains unaccounted for in the data. To get a better understanding of the structure of the data, modify the script so that a flexible spline is fit to the data.

To modify the script and fit a spline, remove the **Fit Line( )** command, and add the **Fit Spline(0.0001)** command:

```
CO2_dt = Open( "$SAMPLE_DATA/Time Series/CO2.jmp" );

CO2_dt << Bivariate( Y( :CO2 ),
       X( :Name( "Year&Month" ) ),
       Fit Spline(0.0001)
);
```

The syntax for the **Fit Spline** command matches the menu option in the Bivariate platform. Because the smoothness of the spline is needed, additional information is included in the parentheses. As you learn JSL, you will find that many commands have the same syntax as they do in JMP menu options.

Figure 1.6 CO2 versus Year&Month Fit Spline

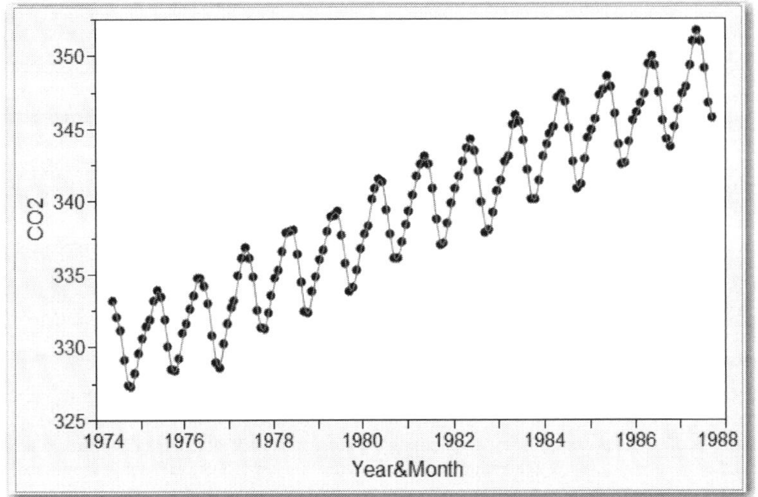

This is an example of scripting, not a proper statistical analysis, so we feel that a brief comment on this example is necessary. The periodicity in the data is obvious—fitting a simple line to the data would usually be insufficient. For an analysis of this data set that supports prediction, the methods in the JMP Time Series platform are needed.

To save the script, select **File ▶ Save**, or select **Save As** and provide a filename such as CO2.jsl. The script is saved as a text file that can be opened by any text editor. If the .jsl extension is used, then JMP recognizes it as a type of JMP file, and will open the file in JMP when it is double-clicked.

To open the script, select **File ▶ Open**, and navigate through the folders to find the script. You can also double-click on a script to open it, or drag and drop the script into another JMP window or into the JMP Home window.

Make It Stop!

As you become more familiar with JSL, and you learn about iterative looping, an important thing to know is how to stop a runaway script. It's not that hard to write a script that goes into an infinite loop that needs to be stopped.

The following script is one that you will certainly want to stop before it gets to the end. To stop a script, select **Edit ▶ Stop Script**. Or, if you are in Windows and the caption is in focus, press the ESC key. Many scripts execute faster than you can stop them. Not this one, however!

```
For( i = 99, i > 0, i--,
Caption( Wait( 2 ), {10, 30},
   Char( i )
   || " bottles of beer on the wall, "
   || Char( i )
   ||" bottles of beer; take one down  pass it around, "
   || Char( i - 1 )
   || " bottles of beer on the wall. "
   )
);
Wait( 3 );
Caption( Remove );
```

Stopping a script introduces the concept of handling the flow of a program. As more advanced topics are discussed, the concept of program flow (i.e., starting and stopping a script, error-checking, and capturing user input) are included.

The Script Window

A Bugatti Veyron is to a car what the JSL Script window is to a text editor. A car gets you where you are going, but a Veyron can get you there much faster. Similarly, the JSL Script window is not just a text editor, its features help you write and debug your script faster.

One of the more useful features of the Script window is the ability to show line numbers to the left of the code (see Figure 1.7). This helps you keep track of progress and debug the script. If line numbers are not showing, then in the Script window, right-click, and select **Show Line Numbers**. Even though the default in JMP is that line numbers are not shown, we recommend that this feature be used.

In the script window, several other features are useful and worth mentioning:

- Text for JMP keywords, strings, comments, and scalar values are color-coded.
- The script can be reformatted for readability.
- JSL functions can be auto-completed.
- If you hover over JSL functions and variables, tooltips are displayed.
- Brace matching is available.

Figure 1.7 Script Window

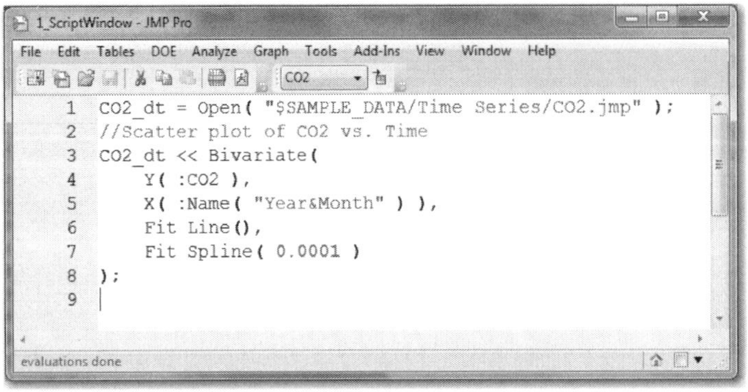
1_ScriptWindow - JMP Pro
File Edit Tables DOE Analyze Graph Tools Add-Ins View Window Help
CO2
CO2_dt = Open("$SAMPLE_DATA/Time Series/CO2.jmp");
//Scatter plot of CO2 vs. Time
CO2_dt << Bivariate(
Y(:CO2),
X(:Name("Year&Month")),
Fit Line(),
Fit Spline(0.0001)
);
evaluations done

Understand the Features of the Script Window

The *JMP Scripting Guide*, included in the JMP installation and available by selecting **Help ► Books**, gives a complete description of the features of the Script window. You are encouraged to refer to the guide often for additional details. A script named JMP Script Editor Tour.jsl is also included. It is available by selecting **Help ► Sample Data**, and clicking **Open the Sample Scripts Folder**.

Color of Code

When you create or open a script, you will notice that certain types of words or text are in different colors to make the script easier to read and debug. If you are familiar with SAS, you will notice that the coloring is similar to SAS code. The colors discussed in this section refer to JMP default colors, which are configurable in the preferences. In the script shown in Figure 1.7 and included in the 1_ScriptWindow.jsl script, the following conventions were used:

- JMP functions, such as **Open( )**, are blue.
- Strings, such as Year&Month, are magenta.
- Comments are green.
- Scalar values are bold and cyan.
- All other text is black.

Reformat Script

Everyone has their own style of spacing and indenting when scripting. It might make perfect sense to the person scripting, but makes no sense to the people who are trying to interpret the code or debug it. The Reformat Script option uses JMP default spacing and indenting to make the script's format standardized and easier to read. When a script window is active, the Reformat Script option can be selected from the Edit menu, or by right-clicking on the script. When this option is run, if there are syntax problems, such as unbalanced parentheses, missing commas, and so on, an error is produced. The script is not reformatted until the syntax errors are fixed, and the Reformat Script option is run again.

Auto-completion of JSL Functions

If you do not remember the exact name of a JSL function, or you are just in a hurry, auto-completion helps you complete the correct syntax of the function. To use auto-completion, type the first few characters, and then hold down the CTRL key, and press the space bar, or hold down the CTRL key, and select the Enter key. As shown in Figure 1.8, if you want to see the properties of a data table, and you have forgotten the syntax, simply type `show`, hold down the CTRL key, and press the space bar. The selection box appears. Select **Show Properties**. Auto-completion can be used after a send operator (**<<**) if the variable to the left of the operator is a reference to an object that accepts messages.

Figure 1.8 Auto-completion

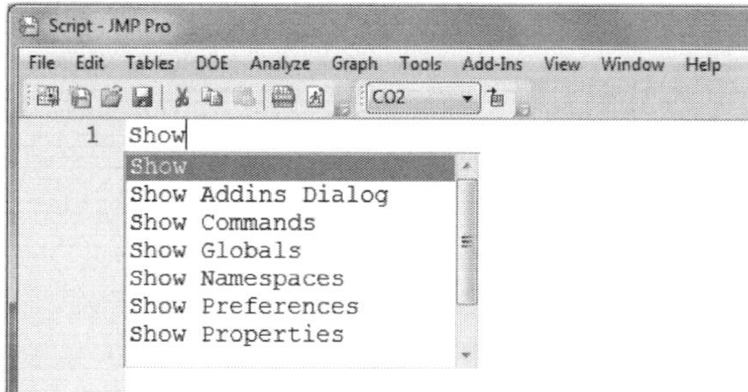

Hovering Over Functions and Variables

In the Script window, when you hover over a JSL function, a tooltip pops up, and shows a brief summary of the syntax. This is extremely useful if you are new to JSL, and you are getting familiar with functions. Hovering over a variable shows a tooltip about the current value of the variable. So, the code needs to have been run before JSL assigns a value to a variable. If the code has not been run successfully, there will be no tooltip when you hover over a variable.

Brace Matching

When we talk about brace matching, we mean matching closing parentheses, brackets, and curly braces with opening ones. There are several facets of this feature.

- When an opening brace is typed, the closing brace is automatically added. If you type the closing brace, JMP recognizes that it has already automatically added the closing one, and does not add the extra one.

- To help check that braces are matched, when you place the cursor on the outside of a brace, its matching brace turns blue. If there is no matching brace, the unmatched brace turns red.
- To select the braces and the text within them, either double-click on a brace, or place your cursor inside the brace, hold down the CTRL key, and select the] key.

Change Script Window Preferences

When you select **File ▶ Preferences**, you can change the preferences in the Script Editor. If the preferences have not been changed since installation, then the script window preferences will look the same as they do in Figure 1.9. The only exception is the Show line numbers option. Even though the default in JMP is that line numbers are not shown, we recommend that this feature be used. This feature helps you debug code because the error message typically includes a line number.

Figure 1.9 Script Window Preferences

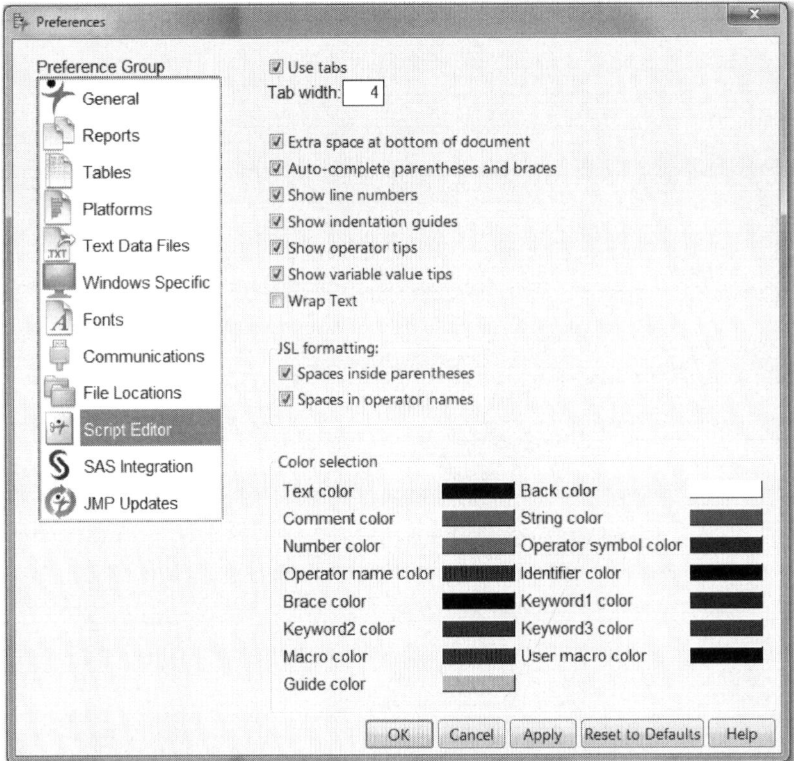

Options can be deselected. However, we have found the default options to be useful, in addition to selecting Show line numbers.

The font used in the script window can be changed. To change the font, select **Fonts** in the **Preference Group**. The **Mono** option controls the font for the Script window.

There a few more items to note about the script window:

- From the **Edit** menu, the **Search** option includes a **Find** (**Replace**) function that supports the use of regular expressions. All of the features in the Search option are available for use in the Script window.
- You can even script the Script window, which is a more advanced topic that is not covered in this section. Information from one script can be captured and written to another script. You can read or write lines of code from one script and store them as a variable to be used later, or you can write them to another script.

The Log Window

When you are scripting, access to the Log window is essential. When a JSL script is run, the Log window captures messages from JMP about the code, errors, and JSL commands and syntax. This information is invaluable as you write scripts. You might want to arrange your windows so that the Script window and the Log window are side by side. This way, you can run portions of the script or the whole script, and immediately check the Log window for errors. The Log window is basically a Script window without line numbers. In fact, JSL code can be executed from the Log window. The Log window is unique in that it captures messages from JMP when the code is run, replicates the executed code, and allows the user to write messages to the Log window. It can also capture messages that will help you write your script.

View the Log Window

If the Log window is not available when JMP is opened, you can open it by selecting **View ► Log**, or by holding down the CTRL key, and selecting the Shift key and L. You can set your preferences so that the Log window appears only when explicitly opened, when text is written to the log, or when JMP is started. If you plan to do a lot of scripting, then setting the Log window preference to open when JMP is started is recommended.

Send Messages to the Log Window

The three functions **Print( )**, **Show( )**, and **Write( )** send messages to the Log window. The **Print( )** function writes text or variable values to the Log window. Each variable value is on a new line, and text is enclosed within double quotation marks. The **Show( )** function is similar to **Print( )**. However, the Show function also includes the variable name, and sets the variable equal to the value. The **Write( )** function is similar to **Print( )**, but it does not enclose text within double quotation marks, and it writes everything on a single line unless a return sequence (\!N) is included.

To show how each of these functions works, run the first two lines of the 1_LogWindow.jsl script, followed by the **Print( )** line, the **Show( )** line, and then the **Write( )** line. Figure 1.10 shows the results. Note the differences between the three functions. The **Show( )** function includes the variable names. The **Write( )** function does not enclose the text within quotation marks, and it writes all of the output on one line.

```
a = 1;
b = "Hi";
Print( a, b );
Show( a, b );
Write( a, b );
```

Figure 1.10 Log Window Output

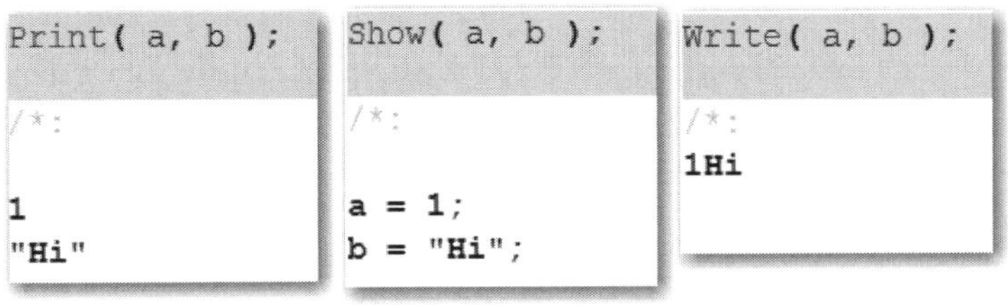

```
Print( a, b );
/*:

1
"Hi"
```

```
Show( a, b );
/*:

a = 1;
b = "Hi";
```

```
Write( a, b );
/*:
1Hi
```

Clear and Save

You will often want to clear the contents of the Log window so that you can see new messages sent to the window. To clear the Log window, right-click in the Log window, and select **Clear Log**. Or, you can select **Select All**, and then select the Delete key. (A keyboard shortcut is to hold down the CTRL and the A keys, and select the Delete key.

If you want to save the contents of a Log window, click on the Log window, and select **File ► Save As**. The default file type is .jsl, and a text file option is available.

Review Error Messages

If there are errors in a script, the messages sent to the Log window will help you debug the code. (There is an entire section in Chapter 10, "Helpful Tips," devoted to debugging code. This section focuses on the output sent to the Log window.) If you run a JSL script with errors, there are three different types of error messages that JMP might produce in the Log window.

1. A JMP Alert. This pop-up window gives a brief message about the type of error encountered, and specifies the line number where it occurred. This type of error halts the execution of the code, and requires the user to click **OK**. The error message in the pop-up window is written to the Log window.
2. The special symbol **/*###*/**. This symbol is embedded in the code that is written to the Log window. The symbol is placed where JMP encounters the error, and an error message precedes the code. We call this "getting pounded."
3. The message **Scriptable[ ]**. This message doesn't always indicate an error, but it is a message that JMP writes to the Log window if there are no syntax errors and no other output produced by the script. This message indicates that the script was executed. It can also indicate that there might be a problem with the code if output was expected.

Example

Figure 1.11 shows the Log window after running the CO2.jsl script. Note how the code is written to the Log window, and how the coloring is retained. The command Bivariate[ ] is printed at the end because it is the result of the executed code.

Figure 1.11 Log Window for the CO2.jsl Script

```
Log - JMP Pro
File  Edit  Tables  DOE  Analyze  Graph  Tools  Add-Ins  View  Window  Help

//:*/
CO2_dt = Open( "$SAMPLE_DATA/Time Series/CO2.jmp" );
//Scatter plot of CO2 vs. Time
CO2_dt << Bivariate(
    Y( :CO2 ),
    X( :Name( "Year&Month" ) ),
    Fit Line(),
    Fit Spline( 0.0001 )
);
/*:

Bivariate[]
```

If the semicolon is omitted from the first line of code, the following error occurs. It suggests what the issue might be, and provides the line number.

Figure 1.12 JMP Alert: Missing Semicolon

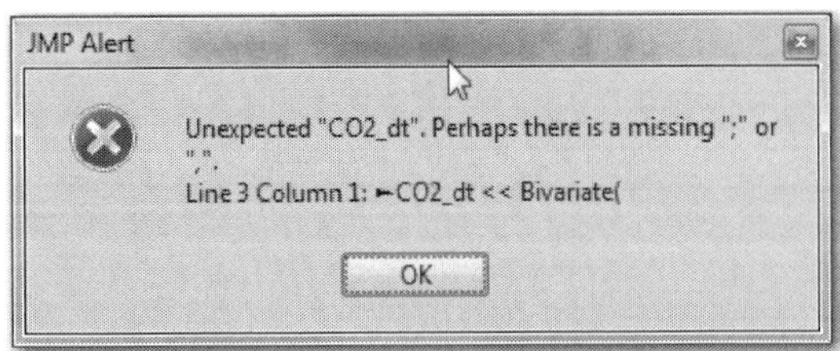

The following description of the error is written in the Log window:

```
Unexpected "CO2_dt". Perhaps there is a missing ";" or ",".
Line 3 Column 1: ►CO2_dt << Bivariate(

The remaining text that was ignored was
CO2_dt<<Bivariate(Y(:CO2),X(:Name"("Year&Month)),Fit Line(),Fit
Spline(0.0001));
```

Suppose that in this script, the keyword Open is spelled incorrectly as Ope. The following error message is sent to the Log window. The error message is not the JMP Alert type—instead, you have been pounded. Note the placement of the special symbol at the end of the line where the misspelled keyword exists, and the error message before the code is replicated.

```
Name Unresolved: Ope in access or evaluation of 'Ope' , Ope(
"$SAMPLE_DATA/Time Series/CO2.jmp" )

In the following script, error marked by /*###*/
CO2_dt = Ope( "$SAMPLE_DATA/Time Series/CO2.jmp" ) /*###*/;
CO2_dt << Bivariate(
Y( :CO2 ),
X( :Name( "Year&Month" ) ),
Fit Line(),
Fit Spline( 0.0001 )
);
```

Get Help with Your Script

This tip might be leaping ahead a bit, but the Get Script command is so useful that we can't resist mentioning it. JMP provides commands that help you write your script by sending the syntax to the Log window. After running the CO2.jsl script, if you run the following command, it produces the code to generate the data file CO2.jmp:

```
Current Data Table() << Get Script;
```

If you run the following code, it lists all of the messages that are available for the data table:

```
Show Properties( Current Data Table() );
```

A Few Items to Note

- The Log window is a script window. Code can be executed from this window.
- When you send the Get Script command to a data table, the Log window captures the syntax of the data table. This will help you write your code.

Let JMP Write Your Script

The most efficient scripter ever on this planet is JMP itself. JMP writes scripts from generated reports or table manipulations. This feature enables a novice scripter to write scripts in a matter of minutes. While teaching an introductory four-hour JSL class, we have seen novice scripters write fairly complex scripts by combining different pieces of code produced by JMP in a script window. Even advanced scripters take advantage of JMP writing their code. It saves them time, ensures that there are no typos, and eliminates the need to search for forgotten syntax.

Capture a JSL Script from a Report

There are numerous ways to capture a script from a JMP report. In addition to capturing the script, you can capture enhancements to the report such as reference lines, changes to the axis scales, inclusions and exclusions of options, and much more. If you click on the top left inverted red triangle in a report window, there is a Script option. If the report produces an analysis using a By Group, then there is also a Script All By-Groups option. Figure 1.13 shows the options available under Script. Only the options directly related to scripting are discussed in this section.

Figure 1.13 Capturing Scripts

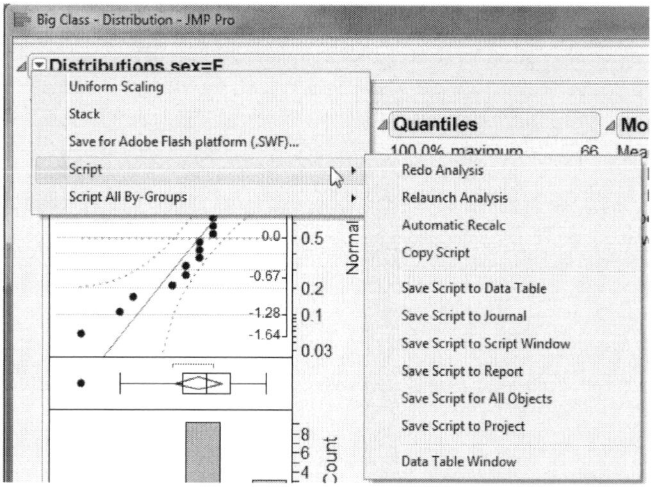

Copy Script—This option copies the script so that it can be pasted into a script window, text file, or any other program that handles text.

Save Script to Data Table—This option saves the script as a table property to the table panel of the data table that generated the report. By default, it gives the table property the name of the platform.

Save Script to Journal—This option creates a link on the current journal, or opens a new journal if one is not open. The link runs a script that reproduces the report.

Save Script to Script Window—This option saves the script for the object to a new script window (if one is not open), or appends it to an open script window.

Save Script to Report—This option writes the script to the top of the report window.

Save Script for All Objects—This option saves the script for all objects in a report to a new script window (if one is not open), or appends it to an open script window. When you save a script for all objects, the **Where** statement defines what is included in a report. It combines all objects in a single window using the **New Window( )** function.

Save Script to Project—This option saves the report window to a JMP project. You can retrieve the script by opening the report window and using one of the other options.

Capture By-Groups Analysis

In addition to the Script option, there might be a Script All By-Groups option. The Script All By-Groups option appears if the report produces a BY-group analysis. The options available in Script All By-Groups are a subset of the options available in Script. The difference between Script and Script All By-Groups is that Script All By-Groups saves the script using the JSL command By, and reproduces the analysis as if you used the By command in a dialog box. With Script, you save the script for all objects, a new window is created, and each object is added to the window.

Capture Table Manipulations

At this point, you know how to save a script from a report that JMP generates. Now, you are going to find out about one of the most powerful and essential features in JMP—its ability to easily manipulate data tables.

When a new data table is generated from a Tables command, the new data table has a table property called Source. The JSL code that generated the new data table from an original table is included in the Source table property. However, there are a few exceptions in which the code is either not captured or not that useful.

Figure 1.14 Source Table Property

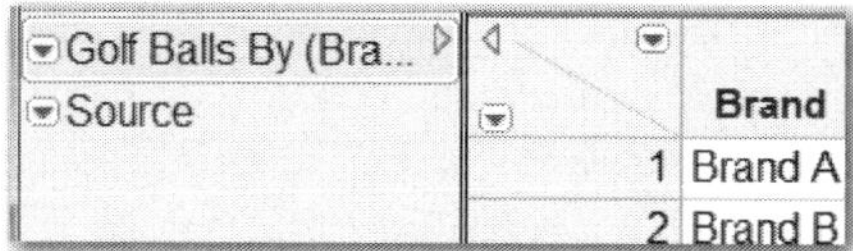

- If you replace a table as a result of selecting **Tables ▶ Sort**, the code is not saved in the Source table property. If you want the code to be saved, do not replace the table when you do the sort. If you do not replace the table, the code is added to the new data table. You can copy and paste the code to another location, and add the option **Replace Table**.
- If you select **Tables ▶ Subset**, the row numbers of the selected rows are included in the code. Having the row numbers is not very useful unless you want the same row numbers every time the code is run. Keep in mind that if you are writing a flexible script, you must select rows and columns before selecting **Tables ▶ Subset**. The commands that select portions of a data table are discussed in detail in Chapter 3, "Modifying and Combining Data Tables."

Example

In this example, the JMP Sample Data file Golf Balls.jmp is used to demonstrate the ability to capture a JSL script from a report and to create a summary table. These two elements are combined in a script window, and they work together to produce the needed output.

Suppose you are asked to examine the distance and durability of different brands of golf balls. You have collected information about three brands. You analyze these brands, but you know that additional brands will be added later, so you want to script a generalized analysis. The three operations required of the script are the following:

1. Create a scatter plot of the relationship between distance and durability. You want to use different colors for each brand to highlight differences in the relationships by brand.
2. Create side-by-side box plots to compare the brands for each response.
3. Create a data table that summarizes the mean and range of distance and durability by brand.

The scripting of these tasks can be accomplished by letting JMP do the work for you! Follow these easy steps:

1. Open the JMP Sample Data file Golf Balls.jmp.
2. In JMP, create a scatter plot of Distance versus Durability. Use the Fit Y by X platform, and add a legend that colors and marks by Brand.

3. Click on the inverted red triangle in the scatter plot, and select **Script ► Save Script to Script Window**.
4. Create multiple box plots in the Fit Y by X platform using Distance and Durability as your Y value, and Brand as your X value. After the box plots are created, click on the inverted red triangle again, and select **Script ► Save Script for All Objects**. The script is saved to the same script window used in the previous step.
5. Create a summary table with the mean and range of Distance and Durability with Brand as the group variable. Click the table property **Source**, and select **Edit**. Right-click on the script, select it, and copy it. Paste the script in the script window used in the previous steps.

This script is now complete. Because this is a simple script that was captured directly from JMP, there are no variable references to tables. As a result, before you run the script, close the summary table that was created. Otherwise, the script can become confused about which data table to use. For your convenience, the script used in the previous example is included for downloading. It is named 1_LetJMPWrite.jsl.

As you script more, you might want to enhance your script. For example, you might want to open the data table directly in the script, reference the data table so that the correct one is always used, and format and save the output. The previous example demonstrated how to write a simple script, but remember you can do so much more!

Get More Help with Your Script

By now, you know that JMP sends valuable information to the Log window. This includes information about a data table generated by the **Get Script** command. The **Get Script** command helps you create a data table by generating the code for adding rows, table variables, columns, formulas, and so on. When you use the **Get Script** command, it prints all of the table values to the Log window, so the output in the Log window can be very long for large tables. A helpful tip for using the **Get Script** command with large tables is to subset the data table to include only the first row of the table in the Log window. The Log window shows the structure of the table, but the length of the output is now shorter and easier to read.

A Few Items to Note

- JMP captures the code required to run analyses or to perform data table manipulations. However, putting the code together in a logical flow, and then adding appropriate references to data tables and reports are critical changes that need to be made to the script for it to run correctly and efficiently.

- JMP captures many items, but it does not do everything. For example, it does not select rows, open tables, reference tables, reformat output, save reports, or save output.

Objects and Messages

In the theater, a script or screenplay is a set of instructions for directors, actors, and staging. In JMP, a script is a set of instructions and commands for creating and manipulating JMP objects. JMP objects include tables, reports, windows, displays, dialog boxes, and much more.

Like in a screenplay, an instruction needs to have a target or a reference. The instruction, "Enter stage right" needs to be targeted for an actor or an object (for example, "Mariachi Band: enter stage right"). Similarly, JSL instructions need a target or reference. Suppose you have the following simple command:

```
Distribution();     //command to open the Distribution platform dialog
```

Note: This command has the same effect as selecting **Analyze ► Distribution** from the JMP main menu.

If you do not have a table open, an Open Data File dialog box appears. After you open the data file, the Distribution dialog box appears.

Open several JMP tables, and run this command. This time, only the Distribution dialog box appears, and Select Columns lists the columns of the active data table. To direct this command to a specific table, a table reference is required.

```
BC_dt = Open( "$SAMPLE_DATA/Big Class.jmp" );
Candy_dt = Open( "$SAMPLE_DATA/Candy Bars.jmp" );

BC_dt << Distribution(); //open Distribution dialog for Big Class
```

Now, let's look at the general syntax of a command:

```
result_reference = object_reference << message(arguments);
```

The result_reference is a variable that is referenced later in the script or JMP session. The object_reference is an object in JMP that can be acted upon, such as a data column, data table, window, or graph. The message(arguments) is a JSL statement (command or instruction). The **<<** is a send operator that sends the message to the object. JMP objects that have built-in commands and properties are described as *scriptable*.

As you script, there are two methods to quickly determine what messages are appropriate for scriptable objects.

The first method is to use JMP online Help. In JMP, select **Help**. In the online Help, you will find the options Statistics Index, JSL Functions, Object Scripting Index, and DisplayBox Scripting Index.

These indexes provide topic help, syntax help, and example scripts ready to run. Figure 1.15 displays the Object Scripting Index for the Bivariate Curve, highlighted in the Objects field on the left. These JMP indexes have saved us countless hours of looking for the correct syntax in the *JMP Scripting Guide*, or searching through numerous project folders for an existing script where a specific command was deployed successfully.

Figure 1.15 Object Scripting Index for the Bivariate Curve

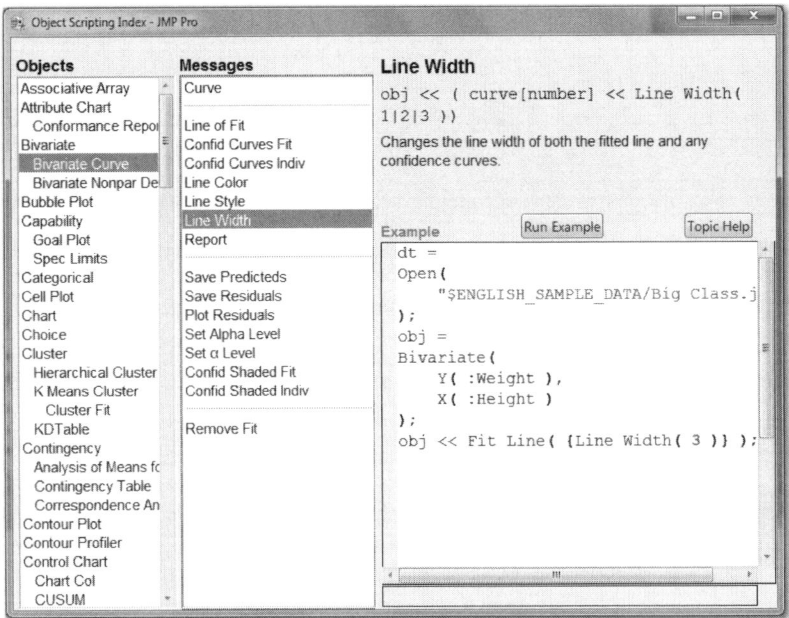

In online Help, the JSL Functions and Object Scripting Index are the most useful indexes for new JSL scripters. The DisplayBox Scripting Index is also useful, but it proves more useful in advanced tasks. Regardless of your experience and knowledge, you should explore the indexes.

The second method is to use the Show Properties(reference) command. This command can be typed in the Log window or in a script window and run. If you type it into the Log window, all messages are listed.

```
Class_dt = Open( "$Sample_Data/Big Class.jmp" );  //table reference
ageCol = Column( class_dt, "age" );  //column reference

//--a1 is the value in the first row of age
a1 = ageCol[1];

//--ageVal is a vector of all values in ageCol
ageVal = ageCol << get values;

//---table is a scriptable object with numerous messages
//--includes Table/Analyze/Graph commands
Show Properties( Class_dt );

//---column is scriptable with many messages
Show Properties( ageCol );

//---a global variable is not scriptable, no messages
Show Properties( a1 );

//---a vector [or a list] is not scriptable,
//---no messages
Show Properties( ageVal );
```

Figure 1.16 Show Properties Output in Log Window

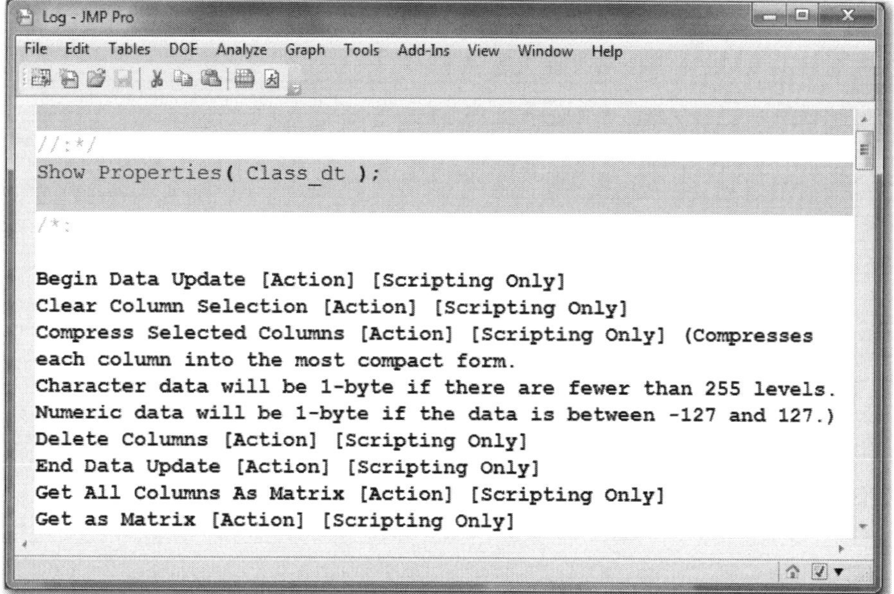

In JMP, *constants* are global variables (such as a1), vectors (such as ageVal), matrices, and lists. Constants are valuable JMP data structures that are not scriptable objects. Constants have no inherent messages. However, each JMP data structure type has a set of functions and lexical rules to create, manipulate, and get information. Notice the long scrollable list of available functions for the matrix data structure in Figure 1.17.

Figure 1.17 JSL Functions Index for Matrix

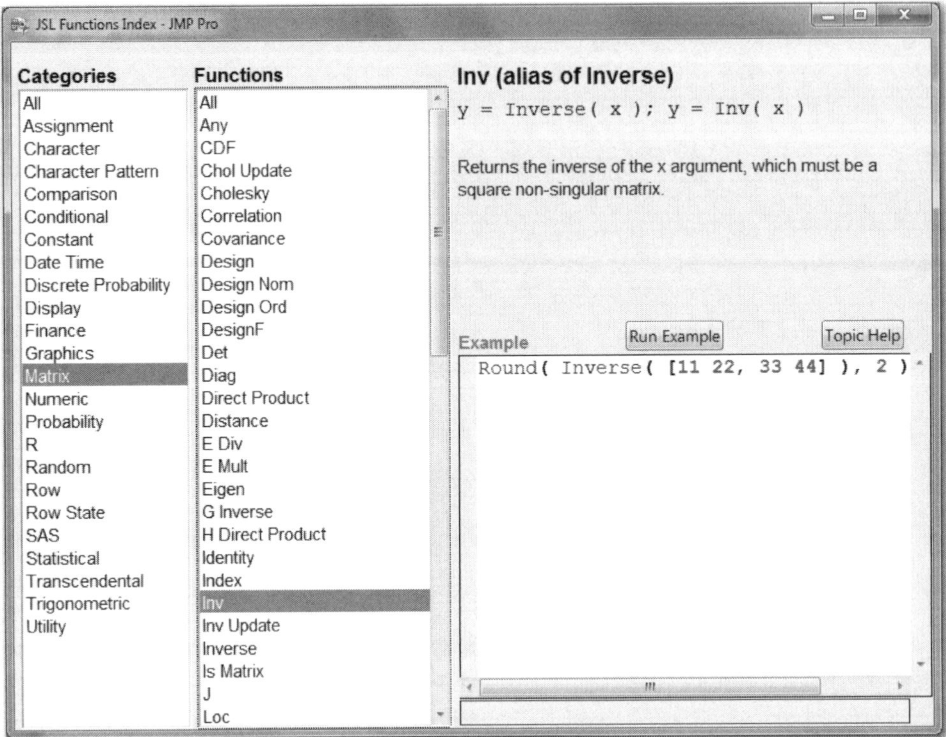

The JSL Functions Index is a superset of categories from the Formula Editor. There are no R, SAS, or Utility categories in the Formula Editor.

You should browse the Utility category, which includes definitions and example scripts for getting information about objects in a script or for communicating with script users. Show Properties, Type, IsTable, Caption, and Dialog are just a few of the functions available in the Utility category.

The 1_ObjectProperties.jsl script includes a list and a vector constant. Both of these are important JMP data structures.

Punctuation and Spacing

In most languages, punctuation can be defined as the use of standard marks in writing to separate words into sentences, clauses, and phrases in order to clarify meaning. Similarly, words in the JMP scripting language are separated by commas, quotation marks, parentheses, semicolons, various operators (such as {}, /, +, –), and so on. It is important to use punctuation properly to clearly express your scripting intentions. In most situations, the existence of a space, tab, or return, inside or between operators or within words, is treated by JMP as if it doesn't exist. However, there are a few situations where one of these *does* matter. This section shows some good and bad examples of punctuation and spacing. The examples are included in the 1_PunctuationSpacing.jsl script.

Use Punctuation

In JSL, commas separate items, such as elements in a list, rows in a matrix, or arguments in a function. Semicolons glue statements together. Curly braces and brackets define lists, subscripts, and matrices. Strings are enclosed in double quotation marks. Parentheses delimit function arguments, and group arguments in an expression.

Consider the following lines of a script:

```
thislist = {3, 7, 31};
thatlist = {"Oregon", "Arizona", "New Mexico"};
thismatrix = [1 2 3, 4 5 6, 7 8 9];
For( i = 1, i < 10, i++,
  Show( i, Factorial( i ) )
);
```

Note how the commas separate the items in the script whether the items are elements in a list, rows in a matrix, or arguments in a function. Each line in the script has a semicolon at the end that glues it to the subsequent line. Lists can be defined by curly braces or with the **List( )** function. Matrices can be defined by square brackets or with the **Matrix** function. Strings are enclosed within double quotation marks. Parentheses tell functions what arguments to evaluate. It is important to follow an opening brace, bracket, double quotation mark, and parenthesis with a closing one to avoid errors or unintended consequences. The JSL script editor can automatically complete incomplete parentheses and braces by selecting the Auto-complete parentheses and braces check box in Preferences. This makes it more difficult (although not impossible!) to make this error. We strongly recommend using this feature.

The script editor can match *fences*, another word for parentheses, brackets, and braces. Hold down the CTRL key, and select the] key in the script. The script editor searches for the first set of fences, highlighting the text between the fences. Repeat this action to highlight the next set of fences.

What if you need to use double quotation marks in a string? JSL provides several escape sequences, and a backslash bang (\!) is the escape sequence to use when you need double quotation marks. For example, if you want the quoted string "Rescue me!" in a **Caption( )**, then your script would look like the following:

```
Caption( "\!"Rescue me!\!"" );
```

Use Spacing

In most situations, JSL doesn't care about a space or tab, or a line or page delimiter in a name. In fact, JMP acts like it doesn't exist. Although there are valid justifications for this behavior, it can be important for a scripter to know.

Consider these lines of the previous script:

```
thatlist={"Oregon", "Arizona", "New Mexico"};
thismatrix = [1 2 3, 4 5 6, 7 8 9];
```

The Log window displays the same list when you run show(thatlist); or show(th at list);, which is not a problem. However, the space between "New" and "Mexico" in "New Mexico" is important because it is part of the string, and for instance, a search for "NewMexico" would not find the desired state name in that list. The columns of the matrix are defined by spaces, and thismatrix has three rows (separated by commas) and three columns (separated by spaces). You get very different matrices with and without spaces and commas!

The two-character operator, such as ||, >=, <=, **, ++, or /*, cannot have a space between the characters to be understood correctly.

You can disable Spaces inside parentheses and Spaces in operator names in the Script Editor Preferences. They are enabled by default. We esthetically prefer the formatted scripts that the default settings generate.

Above all, we encourage you to develop a consistent style in your scripting. Since spaces are important only in a few situations, spacing can be a style element. Spaces can make your scripts more descriptive and easier to read and understand.

Rules for Naming Variables

Simply put, everything you plan on using later in a script needs a name. Relative to some other languages (like SAS, for example), scripts written in JSL have fewer rules for variable naming. However, knowing them and following them can save you a lot of time and effort, not to mention sanity.

As stated in the *JMP Scripting Guide*, the following objects can be assigned a variable name:

- columns and table variables in a data table
- global variables (reference values that can be numbers, strings, lists, and references to objects)
- types of scriptable objects
- parameters and local variables inside formulas

A variable name in a script is resolved the first time it is used. This typically happens when getting or setting a value. The variable value persists forever, or at least until it is deleted or changed, or the JMP session is ended.

All variable naming rules are listed in the *JMP Scripting Guide*. It is a good idea to understand them before getting too far down the road on your JSL journey. This section highlights what we think are the more important naming rules.

First off, scoping is an important concept. Scoping syntax tells JMP how to interpret a variable name in a case that could be ambiguous, such as when you have a data table column and a JSL global variable with the same name. Scoping is described in more detail in Chapter 4, “Essentials: Variables, Formats, and Expressions,” and Chapter 9, “Writing Flexible Code.”

Open the 1_Naming.jsl script, which uses the Sample Data table Body Measurements.jmp. Run the following code:

```
Clear Symbols();  //erases the values set for variables

//Open data table and define Body Measurements.jmp dataset as BMI_dt.
BMI_dt = Open( "$SAMPLE_DATA\Body Measurements.jmp" );

<80Waist_col = New Column("<80 Waist");
```

The data table name BMI_dt resolves without error. However, there is going to be an unexpected < in the Log window. And, the column naming error in the last line is so egregious that JMP displays an Error Alert window.

To fix these problems, either of the following two lines can be used:

```
lt80Waist_col = New Column( "<80 Waist" );
Name( "<80Waist_col" ) = New Column( "<80 Waist" );
```

The first line works because the variable name starts with an alphabetic character. The second line works because of the special parser directive, the Name("...") convention. If the second line of script is run before deleting the column created in the first line, then the column name will be <80 Waist 2 because a column in the table is already named <80 Waist.

The general rule of variable naming in JSL is to start with an alphabetic character or an underscore, unless you use the Name("...") convention. After the letter or underscore, numbers, spaces, and some special characters can be used with abandon. Using the Name("...") convention allows the use of all special characters. You can even use double quotation marks with the backslash bang (\!) operator. Keep in mind that unconventional names can cause problems when exporting data to other formats, particularly when using ODBC.

JSL ignores spaces and tabs in variable names, unless they are enclosed within quotation marks. For example, Var1 2 is equivalent to var12.

There are some reserved words that might cause problems with variable naming. These reserved words are mostly functions. Here's a hint: if the name that you type turns blue (or turns the color that you've chosen as the keyword color in the preferences) when you type it, it might cause a problem. For example:

```
beta = 0.05;
```

In this case, there is a function for a distribution named beta. The variable name can be resolved by using the special parser directive:

```
Name("beta")=0.05;
```

Our advice is to avoid using JMP keywords as variable names.

From the ***JMP Scripting Guide***, JMP has six possible resolutions for a name in a JSL script. In the order in which they are used, they are:

1. Look it up as a function if it is followed by parentheses.
2. Look it up as a global variable unless it is preceded by a single colon. (A single colon indicates a data table column.)
3. Look it up as a table column or table variable unless it is preceded by two colons. (Two colons indicate a global variable.)

4. Look it up as a local variable.
5. Look it up as a platform name.
6. Look it up as an L-value. (An L-value is an assignment address and is explained more in Chapter 4.)

A colon can be used as a scoping operator. For example:

```
::var;  // Var is a Global Variable

:var;// Var is a Table Column

var; // Depends on when first used
```

Capitalization is ignored by JSL. But, that doesn't mean that you should ignore capitalization when you write a script. The use of camel case to name columns can make scripts easier to read and understand. Capitalizing platform names (such as **Distribution**, **Bivariate**, etc.) can make them more obvious and easier to find.

Develop your own consistent style when naming data tables, columns, global variables, etc. Using a standardized style can help a team of script writers generate consistent and understandable scripts that can be seamlessly integrated. More often than not, a consistent naming style can save many hours of frustrating work. Whatever style that you choose, you might be typing the name over and over, so keep the names simple and descriptive. Always use comments liberally in your scripts to help others understand your intentions and logic. You might find that these conventions help you when you revisit scripts that you wrote weeks, months, or years ago.

Operators

Without operators, scripting might read like a novel. Operators get things done! JSL has different types of operators: arithmetic, scoping, datetime, logical, matrix, comparison, and more.

There are three basic categories of operators: prefix, infix, and postfix. As you might guess from the Latin roots of these names, a prefix operator comes before the operand (the object being acted upon); for example, the negative sign (-). An infix operator comes between the operands; for example, the subtraction sign (–). And, a postfix operator comes after the operand; for example, the decrement sign (--). Some operators can be of several types, depending on their use. Operators can be substituted with JSL functions. For example, c = a - b can also be performed with c = Subtract(a, b).

All operators are documented in the *JMP Scripting Guide*. Here are several common operators with script examples and descriptions. These examples are in the 1_Operators.jsl script.

Table 1.1 Common JSL Operators

Operator	Script Example	Description
Arithmetic Operators		
Prefix: **- (unary)**	Result = -a	Result returns the negative of a.
Infix: **+, -, *, /, ^**	Result = a*b/c–d^e	Result returns a times b divided by c, that quantity minus d raised to the power of e.
Postfix: **++, --**	Result = 0; For(a=0, a<=100, a++, Result=Result+a);	Result returns the sum of the numbers from 0 to 100.
Datetime Operators		
Week of Year, Date DMY	Result=Week of Year(Date DMY(20,1,1968));	Result returns 3, the third week of the year, for 20Jan1968.
Assignment Operators		
=	Result=a;	Assigns the current value of a to Result; replaces Result with a.
Comparison/Logical Operators		
==	Result==a;	Boolean logical value for comparisons. Returns 1 if true, 0 if false. Missing values in either Result or a causes a return value of missing. This case evaluates as neither true nor false.
<, <=	a<b	Returns a 1 if a is less than b, a 0 if not; missing values in either a or b return missing.
>, >=	a>=b	Returns a 1 if a is greater than or equal to b, a 0 if not; missing values in either a or b return missing.
& (and)	Result=a & b;	Boolean logical And( ). Returns true if both are true. See the paragraph about behavior for missing values.

(continued)

Table 1.1 Common JSL Operators

Operator	Script Example	Description
| (or)	Result=a | b;	Boolean logical Or( ). Returns true if either or both are true. See the paragraph about behavior for missing values.
IsMissing( )	Result= If (IsMissing(a), b, c);	IsMissing(a) returns a 1 if a is missing, and a 0 if a is not missing.
Other Operators		
List( ) or { }	Result = List(a,b); Result = {a,b};	Lists are containers in which to store different objects. Lists are powerful. They are discussed briefly in this chapter, and are covered more completely in Chapter 5, "Lists, Matrices, and Associative Arrays."
Concat( )	Result=a || b;	Appends b to a.

This table is not a comprehensive list of all of the operators in JMP. On the contrary, there are an infinite number of operators because you have the power to create your own. For example, you could create an operator named Mag that finds the magnitude of a column vector: Mag = function({a},sqrt(a`*a));. The possibilities are endless!

Missing values require special attention. For most logical and comparison operators, any missing values in the calculation return a missing value in the result. In other words, almost all calculations involving a missing value returns a missing value. There are two notable exceptions to this. If one value is missing and another is true, then Or( ) returns true. If one value is missing and another is false, And( ) returns false. Only numeric values are considered missing. A missing character value is considered a string of zero length, not a missing value. For a character variable named Result, checking for a missing Result could be accomplished by returning Result=="", or IsMissing(Result).

Often, successful script writing depends on evaluating operations in a specific order. JSL has a specific precedence of operator evaluation. For example, in the formula d = c * a - b, the expression c * a is evaluated before b is subtracted from the product. It might behoove you to familiarize yourself with precedents in the *JMP Scripting Guide*. The inside-out order of operation can be controlled with an apparent overuse of parentheses. However, some scripting aficionados might consider a plethora of parentheses as gauche and amateurish. The order of evaluation can be surprising, so make sure the operators you use return the results that you intend. Don't be afraid to overuse parentheses!

Lists: A Bridge to Next-Tier Scripting

At this point, you might think that we believe too many characteristics of JSL are *essential.* But, understanding the JMP list object and several list functions is *really essential* for creating scripts that interact with the user and for customizing output. Because lists are so indispensable when writing scripts, this section gives a preview. Additional sections in Chapter 5 provide more in-depth information and examples. As you learn about lists, it might be helpful to read this section first, and then skip to the sections on lists in Chapter 5.

Lists are pervasive in JMP. For example:

- Built-in Dialog boxes save user responses in a list.
- Report windows are lists of display boxes.
- The **Summarize( )** function, that produces results similar to Tables ▶ Summary, stores results in lists and vectors.

Lists are compound data structures for numbers, text, functions, expressions, matrices, and even other lists. The data structure is described as *compound* because it can be a container for other data structures.

Curly braces are the lexical representation for the List function **List( )**. Items in a list are separated by commas.

```
myList = {1, 2, 3};
myFormalList = List( 1, 2, 3 );
Show( myList, myFormalList );   //same result
```

JSL provides many functions to extract information and manipulate lists. These functions are provided in the *JMP Scripting Guide* and in the online Help under JSL Functions, category All or Utility. A few of the more frequently used functions are in the following table. The examples in this section are included in the 1_Lists.jsl script. For the functions in Table 1.2, the lists are defined as follows:

```
A = {"a", "b", "c", "d", "e", "f", "g", "H", "I"};
B = {1, 0, 2, 0, 3, 0, 2, 3, 0, 3};
C = {1, "a", {1, 2, 3}, {{"KAA", "NM"}, {"AMM", "OR"},
   {"JTZ", "NE"}}, [1, 2, 3]};
```

Table 1.2 List Functions for Referencing and Finding Items

Function	Definition
N Items(list **)**	Returns number of items in the list.
[ ] **Subscript(** list, values **)**	References elements in a list. Brackets are also used to define matrices. For lists, brackets are used to reference elements in a list.
Loc(list, value **)**	Returns a matrix (column vector) of locations in the list where the value is found.

```
Show( N Items( A ), N Items( B ), N Items( C ) );
Show( A[3], A[2 :: 4], A[C[1]], C[4], C[4][2][2] );
Show( Subscript( A, 2 ), Subscript( A, C[3] ) );
Show( Loc( B, 2 ), Loc( B, 0 ) );
```

Figure 1.18 Log Window—Lists

```
Log - JMP Pro
File Edit Tables DOE Analyze Graph Tools Add-Ins View Window Help

//:*/
Show( N Items( A ), N Items( B ), N Items( C ) );
/*:

N Items(A) = 9;
N Items(B) = 10;
N Items(C) = 5;
//:*/
Show( A[3], A[2 :: 4], A[C[1]], C[4], C[4][2][2] );
/*:

A[3] = "c";
A[Index(2, 4)] = {"b", "c", "d"};
A[C[1]] = "a";
C[4] = {{"KAA", "NM"}, {"AMM", "OR"}, {"JTZ", "NE"}};
C[4][2][2] = "OR";
//:*/
Show( Subscript( A, 2 ), Subscript( A, C[3] ) );
/*:

A[2] = "b";
A[C[3]] = {"a", "b", "c"};
//:*/
Show( Loc( B, 2 ), Loc( B, 0 ) );
/*:

Loc(B, 2) = [3, 7];
Loc(B, 0) = [2, 4, 6, 9];
```

A Few Items to Note

- Index(2,4) is the function format for 2::4, which is equivalent to [2,3,4]. The Index function allows a third element for increment, which can be negative. Index(6,1,-2) is equivalent to [6,4,2].
- C[4][2][2] is interpreted left to right. C[4] is equivalent to {{"KAA","NM"}, {"AMM", "OR"}, {"JTZ", "NE"}}. C[4][2] is equivalent to {"AMM", "OR"} and C[4][2][2] is "OR". A sequence of {list}[index1][index2]..[indexN] can be indefinitely long.
- **Contains(**list, value**)** returns the first location of the value in the list, or zero if the list does not contain the value. We think the function **Loc(**list, value**)** is more useful, except in the special case where all values in the list are not unique. **Contains( )** returns a scalar, nonnegative integer. **Loc( )** returns a matrix. Loc(A,"c") returns [3] in a 1x1 matrix. **Contains(**A,"c"**)** returns 3.
- In Insert **Into( )** and **Remove From( )**, the double colon (::) before a variable makes the variable a global variable. Global variables are discussed in Chapter 4.

Table 1.3 List Functions for Inserting and Removing Items in a List

Function	Definition
Insert(list, value, ,<i> **)**	Returns a copy of the list with value inserted at the end if a position <i> is not specified. This function can be used to join lists.
Remove(list, i, <n=1> **)** **Remove(** list, {item #s} **)**	Returns a copy of the list with items removed. The starting postion is i. By default, one item is removed. In other words, n=1 if it is not specified.
Insert Into(::x, value, <i> **)**	Inserts value into ::x at position i, or at the end if i is not specified. ::x must be a variable. value can be a single item, a list, or a variable.
Remove From(::x, i, <n=1> **)**	Deletes n items in ::x, starting with position i. If n is not specified, only one item is removed. ::x must be a variable.

```
myAList = {1, 2, 3};
myBList = Insert( myAList, 10 ); //{1, 2, 3, 10}
myCList = Insert( myAList, 10, 2 ); //{1, 10, 2, 3}
myDList = Insert( myAList, {10, 11, 12}, 2 ); //{1, 10, 11, 12, 2, 3}

myEList = Remove( myDList, 2, 2 ); //{1, 12, 2, 3}
```

```
myFList = Remove( myDList, {1, 3, 5} ); //{10, 12, 3}

myAList = {1, 2, 3};
Insert Into( myAList, 10, 2 );    //{1, 10, 2, 3}
Insert Into( myAList, {15, 22} ); //{1, 10, 2, 3, 15, 22}
xx = {-2, -1};
Insert Into( myAList, xx, 1 ); //{-2, -1, 1, 10, 2, 3, 15, 22 }

myAList = {-2, -1, 1, 10, 2, 3, 15, 22};

Remove From( myAList, 2, 2 ); //{-2, 10, 2, 3, 15, 22}

myAList = {-2, -1, 1, 10, 2, 3, 15, 22};
Remove From( myAList, {2, 4, 6, 8} ); //{-2, 1, 2 ,15}
```

Insert( ) and **Remove( )** do not modify the original list. myList=Insert({1,2,3},{4,5,6}) is a valid command. However, **Insert Into( )** and **Remove From( )** have no assignment statement. They are considered in-place commands.

```
myList = Insert( {1, 2, 3}, {4, 5, 6} );  //{1, 2, 3, 4, 5, 6}
Insert Into( {1, 2, 3}, {4, 5, 6} );      //does nothing

ex = {1, 2, 3};
Insert Into( ex, {4, 5, 6} );  //1st argument must be a variable
Show( ex );
```

Insert Into( ) and **Remove From( )** modify the starting list. The first argument must act on a variable (a place to store the results).

Lists in this section have contained numbers and strings, and lists of numbers and strings. Lists can contain expressions, functions, and matrices. Assignment lists and function lists are special cases. In Table 1.4, two functions are listed that are especially useful when working with assignment lists and function lists.

Table 1.4 Functions for Assignment Lists and Function Lists

Function	Definition
Eval List(list **)**	Returns a list where every item is evaluated.
Eval(::x**)**	Eval replaces ::x with its values. Often, this function is applied to a list of column names or references to be used in a command. For example: `Bivariate(Y( eval(yList)) ), X( eval(xList) ) )`

In the following examples, L2 is an assignment list, and L3 is a function list. JMP enables you to reference items in these lists by their "names". For example, L2["x"] is 10. L2[1] is the expression x=10. If you are saying to yourself, "that's not something I'd likely use," put on the brakes. Keep in mind that a JMP dialog box returns a list of user responses in an assignment list. The script in this section includes simple dialog box examples. We are not quite done with this topic. There's more territory to cover regarding lists, expressions, and dialog boxes. But, these few functions should provide you with enough to get started.

For the examples, let:

```
L1 = {1 + 1, Log( 5 ), 1 :: 10, "abc", {10, 20}};  // general list
L2 = {x = 10, y = 1 :: 10, z = 20 * y};     //assignment list

//h function returns value with largest magnitude ignoring sign
h = Function( {x, y}, If( Maximum( x, y ) < 0, Minimum( x, y ),
            Maximum( x, y ) ) );

//g function returns an Empty() if value is +/-9999,missing value
code
g = Function( {x}, If( Abs( Abs( x ) - 9999 ) < .1, Empty(), x ) );

L3 = {h( 2, -3 ), h( -7, -3 ), g( 44 ), g( -9999 ),
   Abs( {44, 25, 9999, -100, -9999, 22} )}; //L3 is a function list

L1Val = Eval List( L1 );
L2Val = Eval List( L2 );
L3Val = Eval List( L3 );
```

Figure 1.19 is an excerpt from the Log window after evaluating each Eval List statement.

Figure 1.19 Log Window Results—Eval List

```
L1Val = Eval List( L1 );
/*:

{2, 1.6094379124341, [1 2 3 4 5 6 7 8 9 10], "abc", {10, 20}}
//:*/
L2Val = Eval List( L2 );
/*:

{10, [1 2 3 4 5 6 7 8 9 10], [20 40 60 80 100 120 140 160 180 200]}
//:*/
L3Val = Eval List( L3 );
/*:

{2, -7, 44, Empty(), {44, 25, 9999, 100, 9999, 22}}
```

A Few Items to Note

- Almost every computer program or language includes data structures like lists, vectors, and matrices. Data structures enable the efficient organization of information, and they are key components for managing large data sets and complex computations. The script in this section introduces the basic syntax for data structures.
- Vectors and matrices store numeric data only. Lists can store numbers, expressions, strings, other lists, and more.

Reading and Saving Data

Introduction

Discovering the stories hidden in your data begins with getting the data. Fortunately, JSL provides multiple methods to read data from many different sources. Common data sources include:

- databases and files created by database query tools
- JMP, SAS, Excel, and simple text files often in .csv or .txt formats
- files generated by equipment or measurement tools

The last source can be simple text files, but some are not so simple. For example, files written in XML format are widespread, and parsing them can be challenging. Often, sources of data, especially equipment data, typically produce one file for each unit processed or each event. Consequently, analyses of multiple events require opening multiple files.

In addition to basic JSL commands to read and save data, Chapter 2 includes a section that demonstrates the **Open Database( )** function, and three sections that handle opening multiple files, messy text, and XML file formats.

Read Data into Data Tables

Most novice scripts either prompt for an existing file or open a known source, with the simplest source being an existing JMP table. A script using the **Open( )** function with an unspecified path enables the user to select from a long list of file types, including JMP tables, text, Excel workbooks, SAS, dBASE, FoxPro, SPSS, and more. Code for this section is included in the 2_ReadData.jsl script.

The general syntax of the **Open( )** function is the function name, followed by the path and filename enclosed within double quotation marks in parentheses. If nothing is specified in the parentheses, then the user is prompted for a file.

```
dt1 = Open();    // user is prompted for file & options
dt2 = Open("$Sample_Data\Solubility.jmp");    // existing JMP table
```

Figure 2.1 Dialog Box for Unspecified Path

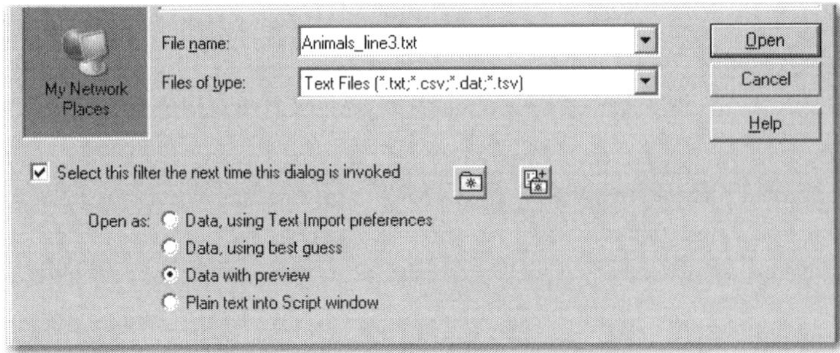

Text Files

To read a text file into a JMP data table, there are several options that handle most simple text files. To see the available options, select **File ► Open ► Files of type ► Text Files**. As shown in Figure 2.1, for the **Open As** option, select **Data with preview**, and then select a file.

Example

Select the Animals_line3.txt file from the JMP folder **Sample Import Data**, and then click **Open**. A window to select available **Import Settings** appears, as shown in Figure 2.2. As the filename suggests, the labels to be used for column headings are on line 3, and the data starts on the next line, line 4. Choose the import settings, and then select **Next ► Import**.

Figure 2.2 Options for Reading Text Files

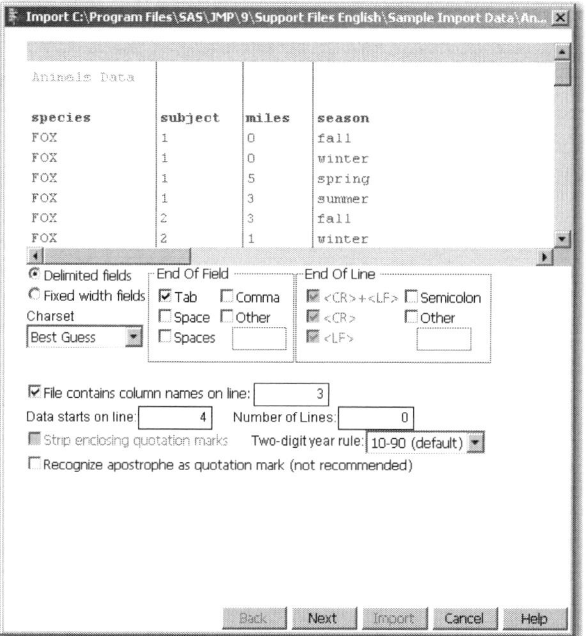

The newly created table has a script named **Source** attached. Select **Edit** from the associated pull-down. The edit window contains JSL similar to the following code:

```
Open( "$Sample_Import_Data\Animals_line3.txt",
  columns( species = Character, subject = Numeric,
           miles = Numeric, season = Character ),
  Import Settings( End Of Line( CRLF, CR, LF ),
  End Of Field( Tab ), Strip Quotes( 1 ),
  Use Apostrophe as Quotation Mark( 0 ), Scan Whole File( 1 ),
  Labels( 1 ), Column Names Start( 3 ), Data Starts( 4 ),
  Lines To Read( All ), Year Rule( "10-90" )));
```

Notice how JMP converts import settings into JSL **Open** arguments. Default options are Charset("Best Guess"), Strip Quotes(1), Labels(1), Column Names Start(1), Data Starts(2), Lines to Read(All). Checked options are Boolean arguments. The 1 value is for checked options, and the 0 value is for unchecked options.

The following command works as well because the defaults apply, except for column names start and data starts:

```
dt3 = Open( "$Sample_Import_Data\Animals_line3.txt",
Import Settings( Column Names Start( 3 ), Data Starts( 4 ) ) );
```

Excel Files

Often, data are stored in Excel. When writing scripts that prompt for the data source, setting the preference **Select Individual Excel Worksheets** is recommended. The command to enable this preference is the following:

```
Set Preference( Select Individual Excel Worksheets( 1 ) );
```

When a user selects an Excel file with multiple worksheets and this preference is enabled, a dialog box prompts the user to select from the list of all worksheets. If more than one worksheet is selected, the last worksheet opened is assigned to the data table reference. The 2_ReadData.jsl script includes an example that prompts the user to select a worksheet using the **Open( )** function. See the section "Retrieve Data from a Database" for more examples of reading Excel files.

Another scripting option that prompts the user and restricts the file type is the **Pick File( )** function, which returns the path of the file. An Open statement is required to read the file into a data table. See the online Help or the *JMP Scripting Guide* for a complete set of **Open( )** arguments. Note that most arguments are file type-specific. Open arguments that are new in JMP 9 are listed in Table 2.1.

Table 2.1 Open Arguments New in JMP 9

Option	File Type	Description
Force Refresh	All files	Closes the specified file if it is already open.
Invisible	All files	Hides the opened table.
Charset( )	Text, HTML	Includes the options Best Guess, utf-8, utf-16, us-ascii, windows-1252, x-mac-roman, x-mac-japanese, shift-jis, euc-jp, utf-16be, and gb2312. For the **Charset( )** option, the arguments must be in double quotation marks.
HTML Table(n)	HTML	Uses the URL as the file path. Specify HTML Table (n) to specify which table on the Web page to import. See the 2_ExtraOpenHTML.jsl script.
Use Labels for VarNames(0|1)	SAS files	Uses SAS labels as JMP columns names. The default value is 0 (false).
flag	All files	May be necessary to open a file without an extension. Includes the options Text, SAS, JMP, Journal, and Script. The documentation of this argument says it must be quoted. However, an unquoted option works. See the 2_ReadData.jsl script for an example.

HTML Tables

The syntax for reading an HTML table is **dt = Open(url, HTML Table (n));**. (See Table 2.1 for a description.) The 2_Extra_OpenHTML.jsl script includes a trivial example. The example is designated as "trivial" because Web page reports are notoriously capricious, and they warrant additional functions to capture errors. This example has no error-checking. See the section "Capturing Errors" in Chapter 9, "Writing Flexible Code," to learn more about the **Try( )** and **Throw( )** functions.

Web reports for current stock prices, weather indicators, factory equipment status, and manufacturing material in progress are commonplace. Imagine the task of creating a daily equipment monitor report for a factory with numerous pieces of equipment. A Web table of equipment status can act as a filter (metadata) for equipment that should be ignored because of scheduled preventive maintenance.

Zipped Files

New in JMP 9 is the ability of the **Open( )** function to read a ZIP archive. Select **Help ▶ Object Scripting Index ▶ Zip Archive** to find available object messages and an example. Usage of this function is restricted to text and blob (binary large object) files. ZIP archives, where each file is a long text string or a blob, often are used instead of databases to exploit ZIP compression. Suppose you have a ZIP file where each record is an MRI (magnetic resonance image), or a file of sensor readout with thousands of results per event. ZIP compression (typically 50-90% or more) is important for storing and subsequently retrieving massive data files.

The 2_Extra_ZipFiles.jsl script shows how to zip and unzip JMP files. In Chapter 9, the section "Calling Other Programs from JSL" describes the Windows **Web( )** function to execute system commands. The 9_Web_DLL.jsl script includes an example to unzip archives.

Set Column Formats

Numeric data often need to be reformatted to produce the appropriate display. JMP offers a wide variety of formats. Figure 2.3 shows the available numeric format types in JMP. Once the general syntax is understood, changing a format using JSL can be straightforward.

Chapter 4, "Essentials: Variables, Formats, and Expressions," includes a separate section, "JMP Dates." This section details formatting and using date and time variables. A special section is needed for this information because date and time are often critical components of an analysis, and they can be confusing with the many format and conversion options that are available.

Figure 2.3 JMP Column Formats—Numeric

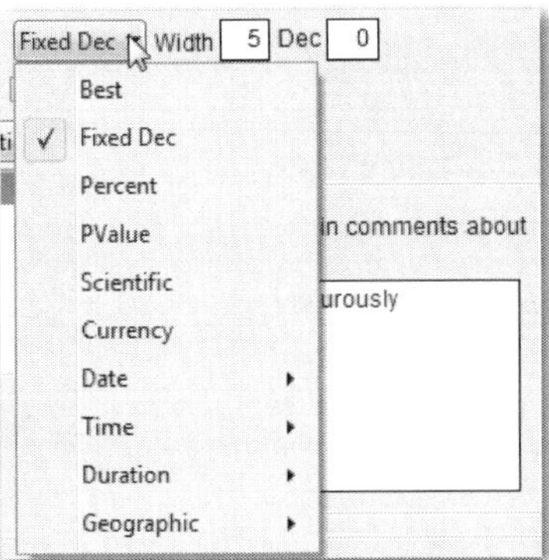

To change the format of a column using JSL, a **Format( )** message is sent to the column or to a column reference. The first argument of the **Format** message is a keyword enclosed within double quotation marks—a string that represents the format type. Assume that the variable col is a column reference in the current data table. Here is the general syntax:

```
col << Format( "type of format", <optional fields> )
```

Figure 2.3 displays the available format types and the keyword arguments for the **Format** message, such as Best or PValue. Format types with multiple options (such as Date) have a drop-down menu symbol (▶). The keyword is defined by the list of options that are displayed when Date is selected. For example, m/d/y is the first option under Date. Other arguments that are available depend on the format that is selected. Table 2.2 lists some formats and their optional fields.

Table 2.2 Examples of Column Formats

Examples	Additional Fields
`col << Format( "Best", 10 );`	Column Width
`col << Format( "Fixed Dec", 10, 2 );`	Column Width, Decimal Places
`col << Format( "Percent", Use thousands separator( 0 ), 10, 2 );`	Use Thousands Separator(0|1), Column Width, Decimal Places
`col << Format( "Pvalue", 10 )`	Column Width
`col << Format( "Scientific", 10 );`	Column Width
`col << Format( "Currency", USD, Use thousands separator( 0 ), 10, 2 );`	Currency Type, Use Thousands Separator(0|1), Column Width, Decimal Places
`col << Format( "m/y", 10, "m/d/y" );`	Display Format, Column Width, Input Format
`col << Format( "m/d/y h:m:s", 19, 1, "m/d/y h:m:s", 1 )`	Display Format, Column Width, Decimal Places, Input Format, Decimal Places
`col << Format( "hr:m:s", 25, -1, "m/d/y h:m:s", 1 )`	Display Format, Column Width, Decimal Places, Input Format, Decimal Places
`col << Format( "Latitude DMS"("PUNDIR"), 15, 0 )`	Geographic Format, Field Punctuation, Direction, Column Width, Decimal Places

There might be instances where the syntax for the column format is not obvious. For example, the syntax for currency formats can be cryptic. JMP uses a three-letter currency standard, but you might need help if you are not familiar with these. The trick to get the syntax for a specific format (or any column property) is to set the format using JMP dialog boxes, and then use the following command:

```
col << get script;
```

When this command is run, JMP writes to the Log window the JSL script that creates that column, including all of its properties and values.

Example

This example uses the 2_ColumnFormats.jsl script. It creates the data table and reformats the columns, as described.

Suppose you work for a toy company, and you are asked to convert the prices of your top-selling toys for infants to different currencies. The data table Currency.jmp, which is created by running the first part of the script, provides a list of your top ten toys, the dates when the prices were set, and the prices in US dollars. There are table variables with currency conversion factors. Columns in the data table use these table variables to convert the prices. This script converts the columns to the appropriate currency formats.

```
//Define data table
currency_dt = Data Table( "Currency.jmp" );

//Update date display to show only month and year
Column( currency_dt, "Date" ) << Format( "m/y", 10, "m/d/y" );

//Update format of currency columns
Column( currency_dt, "Price CAD" ) << Format(
    "Currency", CAD, Use thousands separator( 1 ), 10, 2);
Column( currency_dt, "Price CNY" ) << Format(
    "Currency", CNY, Use thousands separator( 1 ), 10, 2);
Column( currency_dt, "Price NOK" ) << Format(
    "Currency", NOK, Use thousands separator( 1 ), 10, 2);
```

Suppose you do not know that the three-character standard to convert to Norwegian krones is NOK. It is not obvious from the JMP dialog box. Using the method previously described, manually change the column format to Norwegian krones, and then run the following line of code:

```
Column( currency_dt, "Price NOK" ) << get script;
```

The following information is written to the Log window. Note that the format for Norwegian krones is provided. Values are not written to the Log window because this column is based on a formula.

```
New Column( "Price NOK",
  Numeric, Continuous, Format( Currency( "NOK" ), 10, 2 ),
  Formula( :Norwegian Krone * :Price USD ))
```

A Few Items to Note

- To set the format of a column, send a **Format** message to a column. To change the format of a value, use the JSL **Format( )** function and assign it to a variable. Consider the following line of code that uses the **Format( )** function to change the format of the value:

```
val_NOK = Format(187.6091008,"Currency", "NOK", 2 );
```

For timely results, do not use the **Get Script** command for a table or a column with many rows. For columns that are not based on a formula, the values and syntax for the column properties are sent to the Log window. Instead, select a single row of the table, select **Tables ► Subset**, and then run current data table( ) << get script or column reference << get script. This code is included in the App_GetTableScript.jsl script.

Create Data Tables

The data table is the hub for JMP graphs and analyses. It can act as a source (input) and as storage for results (output). Creating a data table is your solution for countless scripting circumstances. Examples include populating a table with known values, formulas, and values stored in matrices or lists. Or, you can populate a table with simulation results. As you script, you will find many reasons to create new data tables.

Here is the general syntax for creating a new data table:

```
new_dt = New Table("table name", <invisible>, <actions> )
```

All arguments are optional. It is possible to use only the **New Table( )** function to create a data table. When no arguments are specified, a table with no rows and named Untitled *n* is created. The *n* represents the nth table created for the current JMP session. It is good programming practice and highly encouraged to give a new data table a descriptive name. To use the invisible option, the table must be assigned a variable name. If the invisible option is specified, the table is not visible to the user or in the list. Creating invisible tables can improve the performance of a script. This feature is discussed more in the section “Performance” in Chapter 10, “Helpful Tips.” Optional arguments include any valid message that can be sent to a data table, such as **Add Rows( )** and **New Column( )**.

Example

Let's start with a simple example that creates a data table, and then populates it with known values. The following code generates a new JMP data table with the name Proportions. This data table contains three columns and eight rows. The three columns include data about height and arm span collected from eight people, and a formula of the ratio of height to arm span. The number of rows to add is not specified. JMP automatically adds the appropriate number of rows based on the number of values. Run the following code to create the new data table. This code is included in the 2_CreateDataTables.jsl script.

```
prop_dt = New Table( "Proportions",
   New Column( "height (inches)", Numeric, Continuous,
       Set Values( [72, 68, 73, 72, 76, 66, 65, 71] )),
   New Column( "arm span (inches)", Numeric, Continuous,
       Set Values( [71, 68, 74, 73, 77, 65, 65, 72] )),
   New Column( "ratio height/arm span", Numeric, Continuous,
       Format( "Fixed Dec", 12, 3 ),
       Formula( :Name( "height (inches)" ) / :Name( "arm span
       (inches)" ) ))
); //end New Table
```

Suppose that you make the following three changes to this data table. You add a continuous color scale (blue to gray to red) for height. You add a note to the column height. And, you change the marker to a P for row 3 to indicate that that row contains Peter's measurements. (See Figure 2.4.)

Figure 2.4 Revised Data Table

		height (inches)	arm span (inches)
•	1	72	71
•	2	68	68
P	3	73	74
•	4	72	73
•	5	76	77
•	6	66	65
•	7	65	65
•	8	71	72

Obviously (because you are reading this book), you want to know how to script changes, rather than manually changing the data table, row, and column properties. The *JMP Scripting Guide* and the JMP online Help, available by selecting **Help ▶ Object Scripting Index ▶ Data Table**, are excellent resources. However, you might not even know what word to look up, or where to get started. The easiest way is to use the **Get Script** command. Run the following line of code, and then review the results in the Log window:

```
prop_dt << get script;
```

The results in the Log window include several options that were not in the original code, such as specifying the number of rows to add and the format of the columns. (See the 2_CreateDataTables.jsl script.) The results include the Notes column property that was added, and a matrix containing information about the row states. The matrix used in the **Set Row States** command is a function of the six row state attributes, with one value for each row. The row state attributes are explained in detail in the section "Row States" in Chapter 3, "Modifying and Combining Data Table."

Add a List or Matrix of Values to a Data Table

Suppose that you have a list of genders for the people in the newly created data table. The following code adds this list to a new column in the data table. The same syntax works if the data are numeric and stored in a matrix:

```
gen = {"M", "F", "M", "M", "M", "F", "F", "F"};
prop_dt << New Column( "gender", Character, Nominal, Values( gen ) );
```

Knowing how to create data tables and controlling the flow of data to tables can improve the performance of your script. This is especially important when handling large data sets or running simulations with many thousands of iterations. In Chapter 10, the section "Performance" provides more details.

A Few Items to Note

If you create an empty JMP table, it will include a column with the name Column 1.

```
JMPtbl_dt = New Table("Empty Table");
```

If you add columns to this table, the column named Column 1 is automatically removed from the table. If you first add rows to this table, Column 1 is retained. See the 2_CreateDataTables.jsl script for an example.

Writing to a data table uses resources. If you want to avoid a table appearing when it is created, you can use the **invisible** option. Remember, if an invisible table is not assigned a variable name, it is not created. Invisible tables do not appear in the list of open tables in JMP. These invisible tables remain open until the JMP session is closed or the table is closed. The following code closes an invisible table using the table's variable name. The function **N Table( )** returns the number of open data tables, which includes invisible tables. The For loop demonstrates a way to close data tables without specifying the table name or the table's variable name.

```
JMPtblinv_dt = New Table( "Invisible Table", invisible );
Close( JMPtblinv_dt, NoSave );

For( i = 1, i <= N Table(), i++,
   nme = Data Table( i ) << get name;
//if needed filter on the name
```

```
Close( Data Table( nme ), NoSave );
//Close( Data Table(i), NoSave) is also valid;
);
```

Another feature in JMP that improves performance is being able to control when the display is updated. The commands **Begin Data Update** and **End Data Update** turn off and turn on the display of table updates. Run the first segment of the following script. Note that the data table display of prop2_dt does not change, yet the valid **Show** command result in the Log window demonstrates that the table does contain column age. Run the last segment of the script. The data table prop2_dt displays all updates after the **Begin Data Update** command is executed.

```
//--------Begin Data Update & End Data Update ----------
prop2_dt << Begin Data Update;  //turn off display updates
New Column( "age",
   Numeric,
   Continuous,
   Format( "Best", 12 ),
   Set Values( [61, 42, 24, 33, 54, 23, 34, 21] )
);
show( prop2_dt:age[ 3 ] );
//note the table stored these values in memory

//----------run to here----------------------
prop2_dt << End Data Update;  //turn on display updates
//column age is now visible, updates are rendered (visible)
```

Close and Save Data Tables

The commands for closing and saving data tables are simple. Having a plan for the output helps you decide which commands and options to use.

A common technique to manage input and output is to create global variables that define data paths. Even if the script eventually prompts you for input or output files, a lot of time is saved by using global variables when testing. Once the application is running, add the commands to prompt for files. To make a script more extensible, use different global variables for the directory paths and the filenames. The script for this section is named 2_CloseSaveDataTables.jsl.

```
//Change the globals to test the script on different input files
::pathOut = "c:\temp\";
::pathIn = "$Sample_Data\";
::fid = "Candy Bars.jmp";
raw_dt = Open( ::pathIn || fid );
//----Do something: get table information or modify the table
nr_dt = N Row( raw_dt );
nc_dt = N Col( raw_dt );
```

```
//save the table in the default path and using default name
raw_dt << Save( );
//save in directory ::pathOut
ttl = raw_dt << get name;
raw_dt << Save( ::pathOut || ttl || ".jmp" ); //JMP file
raw_dt << Save( ::pathOut || ttl || ".xls" ); //Excel file
raw_dt << Save( ::pathOut || ttl || ".csv" );
//comma, space, value file
raw_dt << Save( ::pathOut || ttl || ".txt" );
//text file uses current Export Settings
//----Save and close
Close( raw_dt, Save( ::pathOut || ttl || ".jmp" ) );
//----Close NoSave is commented out since a table can be closed once
//Close( raw_dt, NoSave );
```

Closing and saving a data table as one of the output formats supported in JMP is just that easy. However, when you close a data table, a few details need to be considered.

- Is the table that you are closing linked to an open analysis window or parent table?
- Should a **Wait( )** statement be added to ensure that linked tables and reports are saved before the source data table is closed?

By default, JMP reports (graph and analysis windows) are linked to a data table. Closing a table closes its linked tables and reports. To keep the analyses, be sure to journal the window, save it as a JRP, or save the corresponding script for each linked report before closing its source data table. Also, a short wait (even as brief as a few seconds) can ensure that large linked reports saved to a network location have completed before the parent table is closed.

Linked tables are created from two commands: **Tables ▶ Summary** and **Tables ▶ Subset**. JMP provides options for each of these commands to specify whether the resulting table should be linked. The default for **Summary** is that the resulting table is linked. The default for **Subset** is that it is not linked. These options must be set with the initial command, not with a *post facto* send (<<) operator.

```
//--create linked tables
raw_dt = Open( ::pathIn || fid );
raw_dt << select where( :Brand == "Hershey" );
sub_dt = raw_dt << Subset( All columns, linked );
sum_dt = raw_dt << Summary(
Group( :Brand ),
Mean( :Calories ),
Link to original data table( 1 )
);
Close( raw_dt );  //closes raw_dt, sum_dt, sub_dt
//--create unlinked tables
raw_dt = Open( ::pathIn || fid );
raw_dt << select where( :Brand == "Hershey" );
sub_dt = raw_dt << Subset( All columns );
sum_dt = raw_dt << Summary(
```

```
Group( :Brand ),
Mean( :Calories ),
Link to original data table( 0 )
);
Close( raw_dt );   //closes only raw_dt
```

Saving a text file with Save("filename.txt") uses the user's export preferences. The 2_Extra_SaveText.jsl script creates a report using Tabulate, and saves the report as text. The script includes commands to get the user's export preferences, write out a text file in the appropriate format, and then restore the original settings.

File Requirements

When writing a script that reads or writes data, and that will be used repeatedly, an important planning step is to determine file requirements and naming standards. A few questions to help initiate this planning step are the following:

- Will the resulting files have the same path each time the script is run? Or, will the path be dynamic, based on something like the input data name, the user's name, or the date when the script was run?
- Will users of this script need to pick the saved filename, or just pick a directory path? Or, will the script be scheduled to run with no user interaction?
- Should the results be saved as a JMP file, XLS, text, HTML, or as a combination of formats?
- If there is more than one resulting file, what naming convention should be used to clearly distinguish each result?
- Do the names need to meet other standards? Is there a restriction on spaces, special characters, length, or file extensions? This is especially important if the file has links to Web pages or Microsoft Office SharePoint Server (MOSS) drives, or if the file is used by other applications.
- If unique files are created each time the script is run, how will the files be purged?

There might be other questions to answer, but this should be enough to get you started. JSL functions that can help you address most of these issues are listed in Table 2.3.

Table 2.3 Utility Functions for File Management

Function	Description
Pick Directory(<prompt_string>)	Opens the browse window for the user to select a path.
File Exists(path_file)	Has a Boolean (1|0) result that allows checking before reading or writing.
Files in Directory(dirname, <recursive>)	Returns a list of files. If the recursive option is specified, all files in the subdirectories are returned.
Create Directory(path)	Creates a specified path.
Delete Directory(path)	Deletes a specified path. Be careful with this one.
Creation Date(path)	Checks the creation date of a file or directory. This function is useful to check for data staleness or to purge older versions.
Delete File(filepath)	Enables the deletion of files. This function is used primarily to delete temporary files created by the JSL script or by other programs called by JSL.
Get Current Directory( )	Returns the directory used by File ▶ Open.
Get Environment Variable(env_var)	Returns the user's system ID if username is specified as the environment variable. The user's system ID is typically the user's login name. This function enables you to name user-specific paths for resulting files. Environment variables are case sensitive for the Linux and Mac operating systems.

The included scripts use these simple, but extremely useful functions.

Retrieve Data from a Database

When the term *database* is used, many people think of a very large storage space that contains terabytes of data. In this section, the term *database* is used much more generally. We use it to describe any organized collection of data intended for one or more uses. This implies that text files, Excel spreadsheets, and very large organized computer storage spaces are all databases. Almost all database software comes with an ODBC driver that allows integration between databases and applications. JMP supports ODBC access to SQL databases using JSL.

As anyone who does much data analysis knows well, retrieving and formatting data can take a significant proportion of the total time required to do an analysis. For turnkey applications and for users who need assistance with data extraction, the ability to pull data directly into JMP can result in significant script reliability improvement and in time saved.

The key JSL function to connect to a database is the following:

```
Open Database( )
```

Open Database( ) has three arguments that serve three purposes:

1. Specifies connection information.
2. Lists SQL statements, the file path containing the SQL statements, or the entire database table name.
3. Specifies the output table name.

Here is the general syntax:

```
dt = Open Database (datasourcename | "Connect Dialog",
"Select …" | "SQLFILE=pathname" | tablename,
outputTablename);
```

Teaching SQL is beyond the scope of this book. However, the examples in this section demonstrate some SQL basics. JMP provides a few database files (.dbf files) in its $Sample_Import_Data directory, and an Excel workbook can also function as a database source. The script included for this section uses both database and Excel files. It is named 2_RetrievingDataBase.jsl. The pathnames in the script might be different, depending on the JMP installation directory for the machine being used. As a result, they might need to be updated to work on your system.

This first example uses the database file Bigclass.dbf. It creates an output table named Bigclass ht wt that contains all of the rows and only the two columns that are selected from the original data table. These two columns are Height and Weight.

```
Bigclass_htwt_dt = Open Database(
 "DSN=dBASE Files;
  DBQ= C:\Program Files\SAS\JMP\9\Support Files English\Sample Import
Data\",
"SELECT HEIGHT, WEIGHT FROM Bigclass",
"Bigclass ht wt"
);
```

The next example shows how to import an entire database table by providing the database name—Solubil.dbf—and using SELECT *. The file path must be explicit for the data source name used in the **Open Database( )** function.

```
Solubil_dt = Open Database(
  "DSN=dBASE Files;
  DBQ= C:\Program Files\SAS\JMP\9\Support Files English\Sample Import
Data\",
"SELECT * FROM Solubil",
"Result"
);
```

Read Data from an Excel File with Multiple Worksheets

The Excel 2007 file included for this section is named Basketball Football Sample Data.xlsx. It has two worksheets. The first worksheet is named College Basketball, and the second worksheet is named Football. The two worksheets are created from two sample data files provided by JMP during installation. These data files are named Basketball.jmp and Football.jmp. The following example scripts use only the Football worksheet, but the Basketball worksheet could have been used just as easily. These examples provide some insight into SQL and its power to carve out portions of data from a data source. Before running this script, save Basketball Football Sample Data.xlsx to the path specified or to any path. Update the script to reflect the path where the Excel file is saved.

```
//--Football is one of the worksheet names
//--each worksheet is treated as a database table
football_dt1 = Open Database( "DSN=Excel Files 2007;
  DBQ=Basketball Football Sample Data.xlsx;
  DefaultDir=C:\Program Files\SAS\JMP\9\Support Files
English\Sample Import Data\;",
  "SELECT * FROM `Football$`",
  "Football");

//--select some fields/parameters/columns
football_dt2=Open Database( "DSN=Excel Files 2007;
  DBQ=Basketball Football Sample Data.xlsx;
  DefaultDir=C:\Program Files\SAS\JMP\9\Support Files
English\Sample Import Data\;",
  "SELECT Height, Weight, Fat, Speed, Neck, Position FROM
  `Football$`",
  "Football_SomeFields");

//--select some fields conditionally
football_dt3=Open Database( "DSN=Excel Files 2007;
  DBQ=Basketball Football Sample Data.xlsx;
  DefaultDir=C:\Program Files\SAS\JMP\9\Support Files
English\Sample Import Data\;",
  "SELECT Height, Weight, Fat, Speed, Neck, Position FROM
  `Football$` WHERE Position='qb'",
  "Football_SomeFieldsConditional");
```

A Few Items to Note

The following command sets the preference for showing or hiding ODBC connection information:

```
pref( ODBC Hide Connection String(1) );
```

If the **ODBC Hide** command is turned off (by changing the 1 to a 0), then two table variables are added to the resulting data table. The first variable contains the SQL code, and the second variable contains the JSL code. Both variables are very useful for understanding syntax. If you connect to a secure database that requires a login, the JSL variable might include the user ID and password, so be careful not to compromise your password security. Figure 2.5 shows the table variables when the data file Bigclass.dbf was opened.

Figure 2.5 ODBC Connection

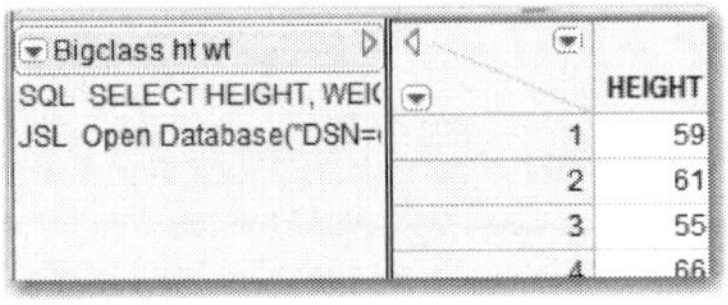

Bigclass ht wt
SQL SELECT HEIGHT, WEI(
JSL Open Database("DSN=

	HEIGHT
1	59
2	61
3	55
4	66

The file path must be explicit for the data source name. A path variable cannot be used. For example, $SAMPLE_IMPORT_DATA/Solubil.dbf is not acceptable. The last set of code in the 2_RetrievingDataBase.jsl script demonstrates how to use character strings with the **Open Database( )** function. This feature makes the query more manageable. It uses functions and syntax introduced in later chapters.

When retrieving data from a specific Excel worksheet, the format that is required is the worksheet name, followed by a dollar sign, and all enclosed within single left quotation marks. The code that creates the data table football_dt1 provides an example:

```
SELECT * FROM 'Football$'
```

When querying for a specific field with spaces or special characters, the field name must be enclosed within single quotation marks. For example:

```
SELECT ID, Parameter, 'Category Value' FROM 'Parameters$'
```

See your local information technologist if you are having difficulties. The database query (DBQ) text might require site-specific, driver-specific, or performance-specific settings, such as DriverId=1046, FIL=excel 12.0, MaxBufferSize=2048, or PageTimeout=5.

Read Multiple Files in a Directory

We are often confronted with the following situation: each time an event occurs, a file is created, and periodically, we need to open and process multiple files for further analyses. Events can be diverse. For example, an event can be:

1. a process recipe is run on a manufacturing tool
2. a customer survey is filled out on a Web site
3. an MRI is taken for a patient

These are just a few of the many possible events that can generate multiple files. Program complexity often scales with the number of files. When processing thousands of files, the simple tasks of selecting a file and saving it have additional concerns. For example, if you are accumulating information, it might be wise to script intermediate saves, instead of one save on completion. Figure 2.6 depicts a high-level punch list for working with multiple files.

Figure 2.6 Multiple File Punch List

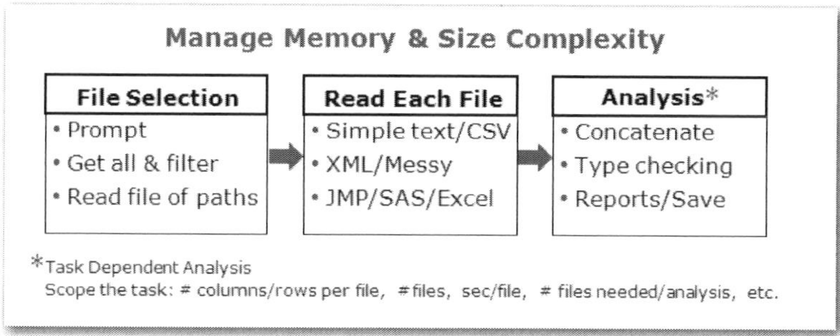

The task of selecting a file is the featured topic for this section. The next two sections tackle the topics of reading XML files and messy text files.

File Selection Functions

The **Pick Directory( )** function initiates a modified browse dialog box that starts at the $HOME directory. This function's only argument is an optional user prompt. It returns the selected path. **Cancel** returns an empty string, and a **Make New Folder** button is available, which is useful when prompting for a directory in which to save files.

```
myPath = Pick Directory( "Select the directory to get all files" );
```

The **Pick File( )** function also initiates a browse dialog box. As of JMP 9, the user can select only one file. This function is useful for an interactive script, where the user must supply an input file for different tasks. For example, an input file can be a configuration file, a spec limits file, or a file of paths to be read and processed (which is pertinent to this section). The function's arguments are listed below. In the online Help, select **JSL Functions Index ► Pick File ► Topic Help** for more details.

```
Pick File(<"caption">, <"initial directory">, < {filter list} >,
< first filter >, < save flag >, < "default file" > );
```

The following code demonstrates the utility of the filter list to restrict the user's selection. This code is a snippet from the 2_ReadData.jsl script.

```
//--Get the directory that is currently the default for opening files.
currDir = Get Current Directory();
//Hover over currDir to see the result

//Note: by using Pick File with only one filter, the user is forced
to select Excel
xlsFID = Pick File(
   " Select an existing Excel file",
   currDir,  //starting directory or leave empty for default
   {"Excel Files|xls; xlsx; xlsm; xlsb"},   //file filter
   1,       //use the first filter
   0,       //0-->Open,  1-->Save
   "Basketball Football Sample Data.xlsx"  //default file
);
```

The function **Files in Directory( )** has two arguments: a valid path and the **recursive** option. This function returns all files, and no filter is available. If the **recursive** option is selected, the files from all subdirectories are returned. Files from the subdirectories include folder names. For example, the Time Series/CO2.jmp file is one item in the sdFids_all_lvl list created by the following commands. The 2_MultipleFileSelection.jsl script includes these commands, and creates a table of filenames from the sdFids_all_lvl list.

```
sdFids_top_lvl = Files in Directory( "$SAMPLE_DATA" );
sdFids_all_lvl = Files in Directory( "$SAMPLE_DATA",recursive );
```

Note: **Pick Directory( )** and **Files in Directory( )** return the filenames in POSIX format, which is the file format preferred by JMP. The **Convert File Path( )** function is demonstrated in the script included for this section. This function is especially useful for applications that use relative paths.

Three Scenarios of File Selection

Typically, system-generated files have a standard convention for the filenames and their directories. Each file has the same general format. But, there are exceptions. Consider the hypothetical MRI event mentioned in the previous section. Suppose each week, a new directory is created and named using the MRI equipment identifier and the workweek. Each MRI file in the directory is named using the same convention: <8-digit patient ID>_<14-digit datetime>_<6-character code for MRI type>, ending with a file extension. The file extension could be custom—.mat, .mri, or .dat. However, the file type is probably XML or ZIP. For example, the 16646646_20111017083027_bilat073.mri file is in the hypothetical directory MRIX2390_201143.

Suppose this multi-file analysis is run with some frequency. Consider three scenarios. In the first scenario, the analysis is interactive. The files are selected by the user running the analyses. In the second scenario, the analysis is semi-automatic. The user specifies a date range or other filter criteria, and the program runs with no further intervention. In the third scenario, the analysis is automatic (or scheduled). By time or by request, the script runs. The script knows the starting path, the dates, and other filter criteria.

Scenario #1: Interactive

- Option A: 1. Use **Pick Directory( )**. 2. Get **Files in Directory ( )**. 3. Create a List Box dialog of files for user to select. 4. Process each file.
- Option B: 1. Use **Pick File( )**, which prompts for a file that lists all the pathnames of the files to be read. 2. Process each file.

Scenario #2: Semi-Automatic

1. Prompt for date and file criteria. 2. Get **Files in Directory ( )**. 3. Filter. 4. Process each file.

Scenario #3: Automatic

1. Use **Files in Directory( )**. 2. Filter by hard-coded or read-in criteria. 3. Process each file.

Another complication is not uncommon. Suppose the naming convention is a randomly generated identifier (which might be expected from a Web survey). Without dates in the name, and no filter in the JMP file selection functions, it might be necessary to open each file in the directory to determine whether it should be processed. The **Creation Date( )** function is handy for this circumstance, and possibly eliminates the need to open each file. The following code is from the 2_MultipleFileSelection.jsl script:

```
sdPath = Convert File Path( "$Sample_Data", Windows);

//-- Get a listing of all the files in the directory & subdirectories
//   since recursive option is specified
sdFids = Files in Directory(sdPath, recursive);

//-- Create a table listing all files
//   Note: the files are alphabetical,
sdFids_dt = New Table("Sample Data Files",
     New Column ("File Paths", character, values( sdFids ))
    );

//-- Creation Date function will return latest date for the file,
//   it can be applied to one file or used in a column formula
sdFids_dt << New Column( "Creation", numeric, Continuous );
```

```
sdFids_dt:Creation << set each value(
   Creation Date( sdPath || :FilePaths ) );
sdFids_dt:Creation << Format( "m/d/y h:m:s", 23, 0 );
```

At this point, you might select files based on the creation date or subfolder, or for a certain string. Run the App_InteractiveMultipleFileSelection.jsl script to see additional features that might be useful for interactive file selection. Figure 2.7 shows this script's user prompts after a directory is chosen. The file selection prompt is dynamic and will not match Figure 2.7. This script uses methods and commands from Chapters 1 through 7.

Figure 2.7 Interactive File Selection (L) Prompt for File Filters (R) File Selection Results

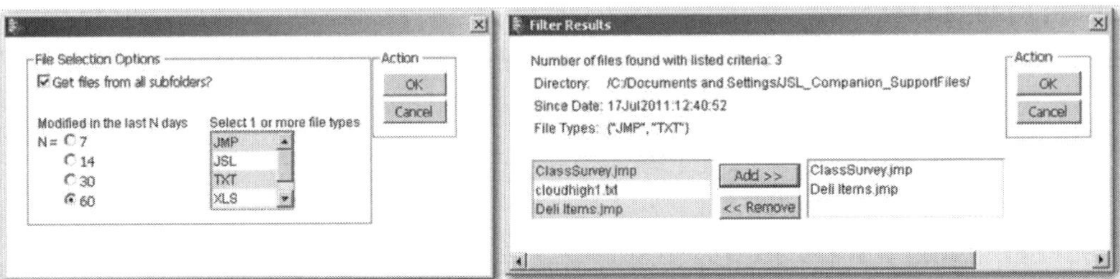

This section and the associated scripts address file selection. As the punch list in Figure 2.6 depicts, there is much more to do. The Read Each File step could be as simple as reading one piece of information, and then storing it in a table. Or, it could be as complex as finding specialized sequences, and then comparing or concatenating all of the information. The file format could be simple, messy, XML, or ZIP, which are covered in other sections. When faced with multi-file analyses, we recommend that you do the following:

- **Scope your analysis.** Ask yourself some questions. Typically, how many files will be processed? What are the average size and the maximum size of a file? How long does it take to process one file? If storing information, how much memory is needed per file? Scoping is like a traveler's budget. It is very frustrating to get to an exotic place, and then find you have no more resources to complete your destination.

- **Manage tables and files.** Do not read the entire file if only a small portion of the information is needed. After processing a file, close it. Reuse global variables if possible, but be careful. Suppose each file being read is assigned to the same variable, dt = open (pathname). Do not reassign the variable until the previous object (table) has been closed. In addition to the **Open( )** and **Close( )** functions, it is worth your time to learn more about **Wait( )**, **Begin Data Update( )**, and **End Data Update( )**. Input and output management and efficient program flow are critical when opening numerous files. Chapter 10 includes functions and suggestions for performance testing.

Commands for File Selection

Operating systems and other programs have commands for file selection that include methods that filter with wildcards and other criteria. Chapter 9 presents the JSL functions **Load DLL( )** and **Web( )** to access external programs. The 2_Extra_MSDirUsingWeb.jsl script provides a simple example and a cautionary note for using **Web( )**, which is a function for Windows only.

Parse Messy Text Files

Messy, like beauty, is a personal assessment. In this book, the adjective *messy* describes any text file that cannot be read into a data table with a single **Open( )** command. It can be a file where custom searches eliminate the need to process many columns and rows. Or, it can be a file that contains different patterns of records.

Messy is the type of file that a PERL script might tackle. If you are a PERL expert, include a **Web( )** command to run a PERL script that converts the messy file into a simple text file, use **Open( )** to read the simplified file, and proceed with your analyses. If you are not a PERL expert, this section provides some JSL commands and tricks to complete the task.

The cloudhigh1.txt file in Figure 2.8 is one of many files that were used in a data visualization competition at the 2006 Joint Statistical Meetings, American Statistical Association Expo. These data were obtained from the NASA Langley Research Center Atmospheric Science Data Center. (See http://isccp.giss.nasa.gov/products/products.html.)

This file represents percent cloud cover. After five lines of file-header information, the data is a 24x24 grid of multiple delimited monthly averages. Columns representing longitudes are in two lines (6 and 7). Latitude is the first field in lines 8 through 31. Latitude is irregular. Some values have multiple spaces, and some do not. This irregularity of columns and spacing meets our definition of messy.

Figure 2.8 File cloudhigh1.txt—Latitude Highlighted

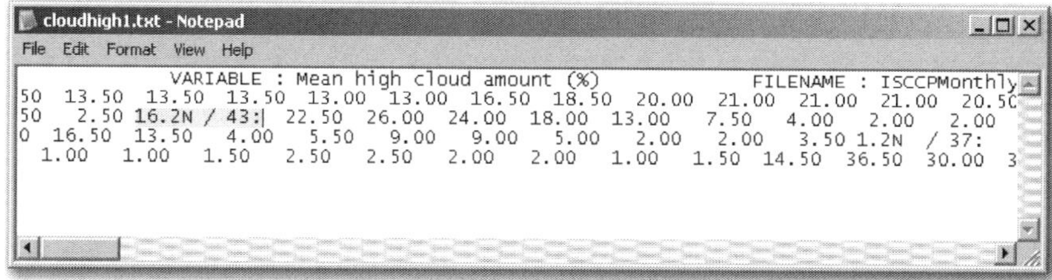

Two-Pass Open Method

If you have information about a file, and it has a known pattern, use it! Recall that **Open( )** options include **Data Starts**, **Lines To Read**, and **End of Field**. These options allow portions of the file to be read.

In the 2_Messy.jsl script, there are two methods for reading the messy text file cloudhigh1.txt. The first method uses two **Open( )** statements. The first **Open( )** statement reads row 8 to the end with no labels. It creates a new column when a colon is found. That first column contains latitude. (See Figure 2.9.)

```
dtdata = Open( jcpath || "cloudhigh1.txt",
   Import Settings(
      End Of Line( CR ),
      End Of Field( Other(":")),
      Labels( 0 ),
      Data Starts( 8 ),
      Lines To Read( All )
   )
);
//--Name column 1 latitude
Column( dtdata, 1 ) << set name( "Latitude" );
```

Figure 2.9 File cloudhigh1.txt—First Pass

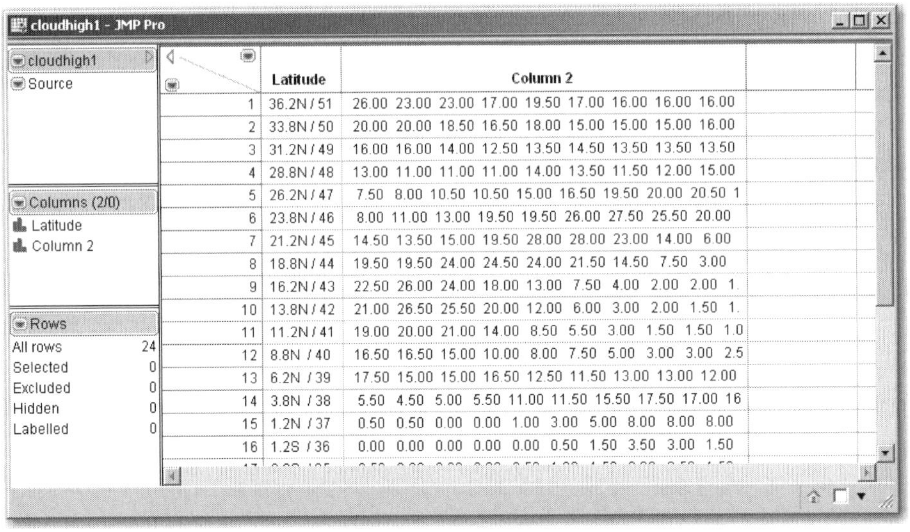

cloudhigh1 - JMP Pro

cloudhigh1
Source

Columns (2/0)
Latitude
Column 2

Rows
All rows 24
Selected 0
Excluded 0
Hidden 0
Labelled 0

	Latitude	Column 2
1	36.2N / 51	26.00 23.00 23.00 17.00 19.50 17.00 16.00 16.00 16.00
2	33.8N / 50	20.00 20.00 18.50 16.50 18.00 15.00 15.00 15.00 16.00
3	31.2N / 49	16.00 16.00 14.00 12.50 13.50 14.50 13.50 13.50 13.50
4	28.8N / 48	13.00 11.00 11.00 11.00 14.00 13.50 11.50 12.00 15.00
5	26.2N / 47	7.50 8.00 10.50 10.50 15.00 16.50 19.50 20.00 20.50 1
6	23.8N / 46	8.00 11.00 13.00 19.50 19.50 26.00 27.50 25.50 20.00
7	21.2N / 45	14.50 13.50 15.00 19.50 28.00 28.00 23.00 14.00 6.00
8	18.8N / 44	19.50 19.50 24.00 24.50 24.00 21.50 14.50 7.50 3.00
9	16.2N / 43	22.50 26.00 24.00 18.00 13.00 7.50 4.00 2.00 2.00 1.
10	13.8N / 42	21.00 26.50 25.50 20.00 12.00 6.00 3.00 2.00 1.50 1.
11	11.2N / 41	19.00 20.00 21.00 14.00 8.50 5.50 3.00 1.50 1.50 1.0
12	8.8N / 40	16.50 16.50 15.00 10.00 8.00 7.50 5.00 3.00 3.00 2.5
13	6.2N / 39	17.50 15.00 15.00 16.50 12.50 11.50 13.00 13.00 12.00
14	3.8N / 38	5.50 4.50 5.00 5.50 11.00 11.50 15.50 17.50 17.00 16
15	1.2N / 37	0.50 0.50 0.00 0.00 1.00 3.00 5.00 8.00 8.00 8.00
16	1.2S / 36	0.00 0.00 0.00 0.00 0.00 0.50 1.50 3.50 3.00 1.50

The second **Open( )** statement reads lines 1 through 7 of the file that contain longitude and other information . This information is read into a single column named Info. (See Figure 2.10.)

```
dtinfo = Open( jcpath || "cloudhigh1.txt",
   columns( Name( "Info" ) = Character ),
   Import Settings(
      End Of Line( CRLF ),
      End Of Field( Tab, Comma ),
      Labels( 0 ),
      Data Starts( 1 ),
      Lines To Read( 7 )
   )
);
```

Figure 2.10 File cloudhigh1.txt—Second Pass

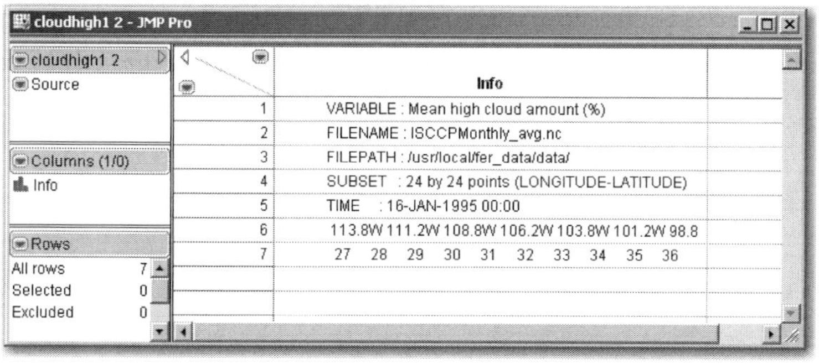
cloudhigh1 2 - JMP Pro
cloudhigh1 2
Source
Columns (1/0)
Info
Rows
All rows 7
Selected 0
Excluded 0
Info
1 VARIABLE : Mean high cloud amount (%)
2 FILENAME : ISCCPMonthly_avg.nc
3 FILEPATH : /usr/local/fer_data/data/
4 SUBSET : 24 by 24 points (LONGITUDE-LATITUDE)
5 TIME : 16-JAN-1995 00:00
6 113.8W 111.2W 108.8W 106.2W 103.8W 101.2W 98.8
7 27 28 29 30 31 32 33 34 35 36

```
//--Process rows 6 & 7, the rows that define the longitude
//--Run to here and hover over lonList1 and lonList2
lonList1 = words(:Info[6]," \!t");
lonList2 = words(:Info[7]," \!t");
```

Combine the data and the information to create the table in Figure 2.11.

```
//--Add 24 columns to dtdata and name them
dtdata << add Multiple Columns("LG",24, After Last, numeric);
For(i=1, i<=24, i++,
//i+2 since the first column has latitude, the data is in column 2
   column(dtdata, i+2) << set name( lonList1[i] || " / "
   ||lonList2[i] )
);

//--Now populate the columns
For( i = 1, i <= 24, i++,
   Column( dtdata, i + 2 ) << set each value(
   Num( Trim( Word( i, Column( dtdata, 2 )[] ) ) )
);
//format each column so it is easier to check
//since this is % cloud cover, 6 wide 2 decimals should be ok
   Column(dtdata, i+2 ) << Format("Fixed Dec", 6,2);
);
```

Figure 2.11 File cloudhigh1.txt—Final JMP Table

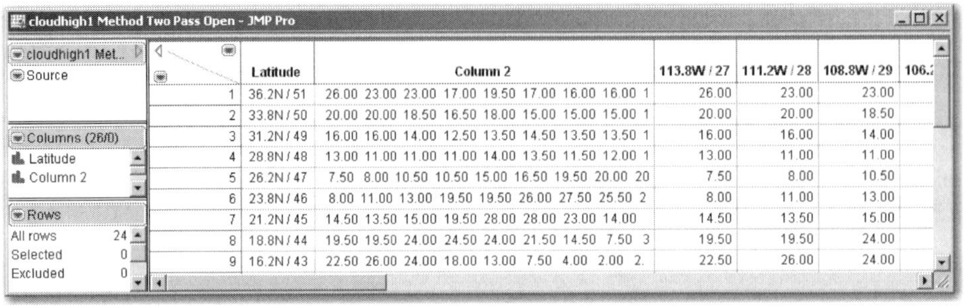

cloudhigh1 Method Two Pass Open - JMP Pro

cloudhigh1 Met...
Source

Columns (26/0)
Latitude
Column 2

Rows
All rows 24
Selected 0
Excluded 0

	Latitude	Column 2	113.8W / 27	111.2W / 28	108.8W / 29	106.:
1	36.2N / 51	26.00 23.00 23.00 17.00 19.50 17.00 16.00 16.00 1	26.00	23.00	23.00	
2	33.8N / 50	20.00 20.00 18.50 16.50 18.00 15.00 15.00 15.00 1	20.00	20.00	18.50	
3	31.2N / 49	16.00 16.00 14.00 12.50 13.50 14.50 13.50 13.50 1	16.00	16.00	14.00	
4	28.8N / 48	13.00 11.00 11.00 11.00 14.00 13.50 11.50 12.00 1	13.00	11.00	11.00	
5	26.2N / 47	7.50 8.00 10.50 10.50 15.00 16.50 19.50 20.00 20	7.50	8.00	10.50	
6	23.8N / 46	8.00 11.00 13.00 19.50 19.50 26.00 27.50 25.50 2	8.00	11.00	13.00	
7	21.2N / 45	14.50 13.50 15.00 19.50 28.00 28.00 23.00 14.00	14.50	13.50	15.00	
8	18.8N / 44	19.50 19.50 24.00 24.50 24.00 21.50 14.50 7.50 3	19.50	19.50	24.00	
9	16.2N / 43	22.50 26.00 24.00 18.00 13.00 7.50 4.00 2.00 2.	22.50	26.00	24.00	

The second method uses **Load Text File( )**, a function that reads the entire text file into a variable instead of a table. This method is described in the next section.

Load Text File

Once the cloudhigh1.txt file is read and assigned to a text variable named fidtext, apply the **Words( )** function, and use a new line and carriage return as delimiters. The result is that each line of the file is now an item in a list.

```
fidtext = Load Text File( jcpath || "cloudhigh1.txt" );
//--Words() partitions a string by delimiters
//--Use newline (\!n) and carriage return (\!r)
fidlines = Words( fidtext, "\!n\!r" );
Show( N Items( fidlines ) );   //31 text strings
```

The parsing techniques after this step are similar to those previously used, and they are not repeated here. The 2_Messy.jsl script has numerous comments to help you comprehend the steps.

Noteworthy items from the script include the following:

1. Information from lines 1 through 5 is used to create a descriptive name and the table property named Notes.

2. Data is populated by fidlines. Data is populated rowwise with nested **For** loops versus the column formula method (columnwise).

3. The code is dense. Techniques discussed in later chapters suggest adding **Show( )** statements or copying several statements into a new Script window. This helps you see what they do and debug. The following snippet is included in the script so that you can look at the results in the Log window and understand what is happening in the nested loops:

```
i=1; j=1; crow=1;
show(fidlines[crow+7]);
tt = Words( fidlines[crow + 7], ":" );
show("step1- get latitude", tt, tt[1], tt[2]);
tt = Words( Trim( tt[2] ), " \!t" ); //tt is repurposed
show ("step2 - get values", tt);
```

Parse with Patterns

Chapter 9 has a section on regular expressions. The 2_Extra_MessyPatterns.jsl script is an extra script. It creates regular expression patterns to read the cloudhigh1.txt file and to create a stacked table of values. It is included as an alternative for reading messy files.

Parse XML Files

JSL has several utility functions to read and write XML. Typically, XML parsing functions are needed to read information encoded on some advanced Web pages. Read-out data from MRIs, satellite photos, earthquake sensors, and manufacturing equipment controllers are often written as text files using XML schema. This section begins with a simple introduction to XML.

XML is a three-letter abbreviation for Extensible Markup Language. Its general structure looks like HTML, which is a coding language for Web displays. Both HTML and XML are used for Web data, but there are marked differences. HTML's goal is to display data, and XML's goal is to describe data. HTML documents only can use tags defined by the standard being deployed on that Web page. XML has no standard tags—the tags and structure must be defined by the author.

Figure 2.12 XML Standard Schema

```
<parent>
  <child>
    <subchild>
    </subchild>
  </child>
</parent>
```

```
<input name="InData" label="Input Data Set" required="true">

<description>Full path and name to the input data. The name must be
lowercase.</description>

</input>
```

All XML documents must contain a single tag pair to define the parent element. All other elements must be nested within the parent element. All elements must be in pairs and correctly nested. Like HTML, XML can have attributes. Attributes must be in the form of name="value" pairs. It is important to note that the attribute value must be enclosed within quotation marks. For the snippet of XML shown in Figure 2.12, input is the parent element; description is its child; and name, label, and required are attributes of the parent element, input.

JMP XML functions are found in the **JSL Functions Index** in the **Utility** category. Here is the typical usage flow:

1. Load an XML file.
2. Assign the contents of the file to a text string.
3. Use **Parse XML( )** to assign values.

The **Parse XML( )** function's first input argument is a string, followed by multiple **On Element( )** arguments. Each **On Element( )** argument has a tag name and two functions, **Start Tag(expr)** and **End Tag(expr)**. The associated expressions for these two functions direct what should be done when the tag is found (**Start Tag**), and how to process the body of XML for this tag name (**End Tag**). Most often, **End Tag(expr)** uses the **XML Text( )** command that formats the XML body as a string, and uses other commands that direct how that string is used.

```
Parse XML(string, On Element("tagname", Start Tag(expr), End
Tag(expr)))
```

The 2_XML_parse.jsl script is a simple example that converts an XML table to a JMP table.

Here is a portion of a simple XML file named CDcatalog.xml. This file contains music CD information. Notice the patterns—**CATALOG** is the parent, and **CD** is a child. Each tag has a closing tag, </tagname>.

```
<?xml version="1.0" encoding="ISO-8859-1"?>
<CATALOG>
  <CD>
     <TITLE>When a Man Loves a Woman</TITLE>
     <ARTIST>Percy Sledge</ARTIST>
     <GENRE>R&amp;B</GENRE>
     <COUNTRY>USA</COUNTRY>
     <COMPANY>Atlantic</COMPANY>
     <PRICE>8.70</PRICE>
     <YEAR>1987</YEAR>
  </CD>
  <CD>
     <TITLE>1999 Grammy Nominees</TITLE>
     <ARTIST>Many</ARTIST>
     ...
  </CD>
</CATALOG>
```

The 2_XML_Parse.jsl script follows the typical usage flow. **Load Text File** reads the file into memory, and then assigns the text to the variable cd_file_contents. Hover over cd_file_contents after the second line of the script is run to see the format. Or, display it in a **Text Box** as shown in the last line of code.

```
fid = "CDcatalog.xml";
cd_file_contents = Load Text File( path || fid);
Parse XML( cd_file_contents,
OnElement("CATALOG",
Start Tag(
   dt = New Table( "CD CATALOG",
   New Column( "TITLE", character ),
   New Column( "ARTIST", character ),
   New Column( "GENRE", character ),
   New Column( "COUNTRY", character ),
   New Column( "COMPANY", character ),
   New Column( "PRICE", numeric ),
   New Column( "YEAR", numeric )
); //end New Table
row=1;)),     //end Start Tag, CATALOG OnElement
OnElement("CD",
   Start Tag( if(row>0, dt << Add Rows( 1 )); ),
   End Tag(row++;)),
OnElement("TITLE", End Tag(column(dt,"TITLE")[Row()] = XML text())),
OnElement("ARTIST", End Tag(column(dt,"ARTIST")[Row()]=
   XMLDecode(XML text()))),
OnElement("GENRE", End Tag(column(dt,"GENRE")[Row()] =
   XML Decode(XML text()))),
OnElement("COUNTRY", End Tag(column(dt,"COUNTRY")[Row()] =
   XML text())),
OnElement("COMPANY", End Tag(column(dt,"COMPANY")[Row()] =
   XML text())),
OnElement("PRICE", End Tag(column(dt,"PRICE")[Row()] =
   num( XML text()))),

OnElement("YEAR", End Tag(column(dt,"YEAR")[Row()] =
   num( XML text())))
); //end Parse XML
New Window("CD XML", Text Box(cd_file_contents));
```

For this example, when the parent element CATALOG is found, table and column headings are created. Each row is nested within the CD tag. **XML text( )** returns the body (the string) between each Start Tag and End Tag.

The **Num( )** character function is used to convert PRICE and YEAR text results to a number. This XML uses an encoding standard for special characters. For any tag that might contain special characters, use the utility function **XML Decode( )** to convert it to a familiar text string. The script uses it for ARTIST and GENRE. It might be prudent to use it for TITLE as well.

For example, the ampersand (&) in R&B is a special character. The XML standard is &. Greater than (>) is >, and less than (<) is <. The 2_XML_Parse.jsl script uses the function **XML Encode( )** to convert a string of special characters to XML, and very few change.

Write XML

The 2_Extra_WriteTableAsXML.jsl script is a concept program. Most of us will never need to write data in XML format. However, it is an opportunity to write a script for a specific task, and it is related to this section's topic. If you are new to JSL, you might want to postpone reading this script until you have mastered Chapters 3 through 5.

3 Modifying and Combining Data Tables

Introduction

Before analyzing any data, the data often need to be cleansed and restructured for the analysis. This is a critical and time-consuming step. There are estimates that as much as 80% of the time spent on analyses is used for this first step. Fortunately, one of the more powerful and useful aspects of JMP is its ability to manipulate and restructure data. With JSL, it is possible to script these time-consuming tasks, and reduce the required effort spent on this step of the analysis. This chapter provides an overview of the most common tasks used to prepare data stored in a JMP data table.

Manipulate Rows

This could be a very long section if we tried to enumerate scenarios requiring row commands. Instead, consider the most common tasks for data tables involving rows: add, delete, or reorder observations; find and change values; and hide or exclude observations from analyses.

As you know from your own JMP experience, row selection is an important and powerful JMP feature. Knowing how to select rows and reference rows are required skills for scripting, and for subsequent sections in this chapter. The snippets of code in this section are all from the 3_ManipulatingRows.jsl script.

Select and Reference Rows

Open the Gosset's Corn.jmp data table from the Sample Data directory, and create a reference to it.

```
//Reference data table
GossetsCorn_dt = Current Data Table();
```

Select specific rows (in this case, rows 2, 4, 6, and 8), and then clear the selections.

Figure 3.1 Gosset's Corn

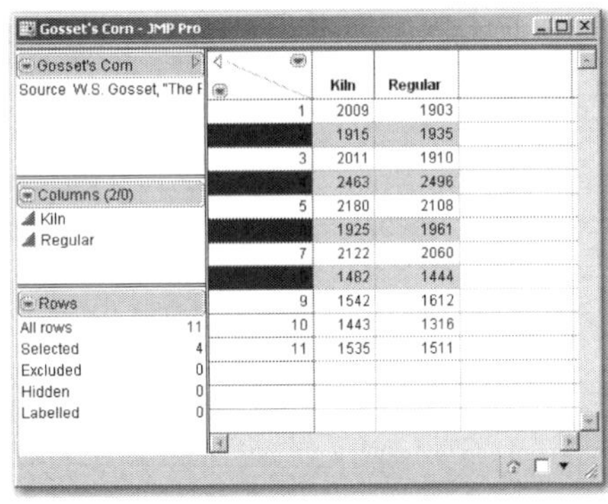


```
//Select four specific rows
GossetsCorn_dt << Select Rows( {2, 4, 6, 8} );

//Clear the selection of rows
GossetsCorn_dt << Clear Select;
```

Find rows with the specified condition, create a vector of row numbers, and then select the rows using the vector.

```
RowChoice = GossetsCorn_dt << Get Rows Where( :KILN >= 1800 );
GossetsCorn_dt << Select Rows( RowChoice );
```

The same conditional selection can be performed with one line of code using the **Select Where( )** command.

```
GossetsCorn_dt << Select Where( :KILN >= 1800 );
```

Select odd-numbered rows, invert the selection to get even-numbered rows, and then delete the rows.

```
GossetsCorn_dt << Select Where( Mod( Row(), 2 ) == 1 );
GossetsCorn_dt << Invert Row Selection();
GossetsCorn_dt << Delete Rows();
```

An alternative method is to select the even-numbered rows initially, eliminating the need (and the demonstration!) of the **Invert Row Selection( )** command.

```
GossetsCorn_dt << Select Where( Mod( Row(), 2 ) == 0 );
```

Add three rows at the beginning of a data table, at the end, and then after row number 5.

```
GossetsCorn_dt << Add Rows( 3, 0 );  //add at the beginning
GossetsCorn_dt << Add Rows( 3 );     //add at the end
GossetsCorn_dt << Add Rows( 3, 5 );  //add after row 5
```

Specific row numbers, lists, and matrices can be used to manipulate rows. This time, an index vector is used to demonstrate a deletion.

```
GossetsCorn_dt << Delete Rows( 1 :: 10 );
//Note: 1::10 = Index(1,10)  = [1 2 3 4 5 6 7 8 9 10]
```

Row commands can be strung together by any number of **Send** statements. This is not the most efficient way to hide and exclude rows where KILN is less than 1800, but it demonstrates this JSL feature.

```
GossetsCorn_dt << Select Where( :KILN >= 1800 ) << Invert Row
Selection() << Hide( 1 ) << Exclude( 1 ) << Clear Select;
```

Assign Values to Rows

The following script creates a new data table with six states from the southwestern United States. This is an example of assigning values for an entire column.

```
States_dt = New Table( "State Data", Add Rows( 6 ),
   New Column( "States", Character, Nominal,
     Set Values( {"TEXAS", "OKLAHOMA", "NEW MEXICO",
      "COLORADO", "ARIZONA", "UTAH"} ))
);
```

An extremely useful function in JSL for manipulating rows is **For Each Row( )**. The following lines of code use **For Each Row( )** to assign values from a list rowwise, and to modify the capitalization of the state names. This is a simple alternative to a **For( )** loop to cycle through rows and change values.

```
stateList = {"texas", "OKlahomA", "New Mexico", "COLORADO",
   "ARIZONA", "utah"} ;
States_dt2 = New Table( "State Data #2", Add Rows( 6 ),
   New Column( "States", Character, Nominal ) );
For Each Row( :States = stateList[ Row()]  );  //assign rowwise

For Each Row( :States = Lowercase( :States ) );
For Each Row( :States = Uppercase( :States ) );
```

A Few Items to Note

- **For Each Row( )** can save you a lot of work! **Row( )** represents the current row number.
- **Select Where(**logical test**)** is a table message to find rows that satisfy the logical test, and it highlights the rows in the data table. **Get Rows Where(**logical test**)** is a function to find and return a vector of row numbers that satisfy the logical test. It does not modify the source table's selections.
- Row commands can be executed individually or stacked sequentially.
- There are often several ways to write a script to get the same result when manipulating rows. The method that you use depends on your programming experience, style, and sophistication.

Row States

A *row state* is a collection of six attributes. Each row in a data table has a value—a single number—representing its *combined row state.* JMP uses row states as specifications about whether and how data appears in graphs and reports. Table 3.1 describes these six attributes.

Table 3.1 Row State Messages in JMP

Messages	Description
Exclude or Unexclude	Excluded rows are omitted from calculations for statistical analyses. Rows that are Excluded show up only in graphs and plots, but not in the summary statistics. Ensure that this is what you intended.
Hide or Unhide	Hidden rows are omitted from graphs and plots. Rows that are Hidden are included only in calculations and statistical analyses. Ensure that this is what you intended.
Label or Unlabel	When rows are Labeled, JMP places row numbers or the values from a designated Label column on points in graphics platforms.
Colors	When rows have colors, JMP uses those colors to distinguish the points in graphics platforms. There are 85 stock color values available without manipulating RGB or HLS values. Custom colors are available by manipulating RGB or HLS values.
Markers	When rows have markers, JMP uses those markers to distinguish the points in graphics platforms.
Selected	When rows are selected, JMP highlights the corresponding points and bars in graphics platforms.

Row states are a key component of interactive JMP graphics. They are easy to manipulate manually in JMP. Using JSL, row state attributes can be modified with messages such as **Hide** or **Label**. Row states can be represented numerically, which is demonstrated later in this section. And, numbers can be converted into row states. They can also be modified arithmetically, which makes scripting row-state management more flexible.

The script for this section is 3_RowStates.jsl. Open the Cars.jmp data table from the Sample Data directory.

Toggle Row States

```
Cars_dt = Current Data Table();
Cars_dt << Select Where( :Doors == 2 );
Cars_dt << Hide;  // Only Hide
```

The script above selects two-door cars from the data table, and hides them in subsequent graphs. Two-door cars are still represented in calculations and reports. The line of code that hides the selected rows acts as a toggle. If those rows were already hidden, and the same code was run, the rows would be unhidden. In the case where some selected rows have the row state set to hide already, and others do not, the line of code first makes all row states consistent, and then, if the same code is run again, it acts as a toggle.

Use caution when setting row states in tables that have previously defined row states. This is especially true for **Exclude** and **Hide**. Even for your own personal applications, it is often helpful to clear the row states before modifying them. We highly recommend this practice when writing scripts that will be used by others. This section's script, 3_RowSates.jsl, contains commands to save a data table's row states before clearing them.

If you want to hide and exclude the two-door cars from calculations and reports, submit the following line of code:

```
Cars_dt << Hide << Exclude;   //Hide and Exclude
```

Multiple row state commands can be executed in a single line of code. It is important to understand the toggling nature of certain row states. For example, if the two previous lines of code commented with `Only Hide` and `Hide and Exclude` are run in sequence, the second **Hide** statement toggles off the first **Hide** statement. This behavior can cause unintended consequences. Be careful with multiple row state commands. Remember, it is often useful to clear the row states when working with a new data table.

```
Cars_dt << Clear Row States;
```

The previous command clears all six row state attributes for each row in the Cars_dt data table.

Get and Set Row States

To get a row state, place the **Row State( )** expression on the right side of the assignment.

```
rs = Row State( 6 );
```

The script creates a variable named rs that stores the row state of row 6. The rs variable appears in the Log window as:

```
Combine States( Selected State( 1 ), Excluded State( 1 ), Hidden
State( 1 ) )
```

To set a row state, place the **Row State( )** expression on the left side of the assignment.

```
Color Of( Row State( 6 ) ) = 3;   //red
```

The script sets the color state of row 6 to value 3, which is red.

A frequently used and handy approach that differentiates the levels of a variable is using **Color By Column( )** and **Marker By Column( )**.

```
Cars_dt << Color By Column( :Doors ) << Marker By Column( :Make );
```

This approach is quick and easy, but it has less specific control over color and marker states than explicitly setting them.

```
Cars_dt << Select Where( Make == "Chrysler" ) << Colors( 14 )
    << Markers( 169 ) << Clear Select;
```

The previous script provides explicit control over color and marker states. It colors rows magenta where the make is defined as Chrysler, and adds the © marker. It might take quite a bit of work to specifically control every category. You need to decide which approach is more efficient and appropriate for your application.

The six row state attributes are packed together in a single number. To see the numeric representation for the row states of row 6 of the Cars.jmp data set after all we've done to it, type the following into the Script window:

```
rs = Row State( 6 );
rscode = Selected( rs ) + 2 * Excluded( rs ) + 4 * Hidden( rs ) + 8 *
Labeled( rs ) + 16 * Marker Of( rs ) + 256 * Color Of( rs );
```

If you executed the script in order, the Log window displays 833 as the value of rscode. Row 6 was selected, marked with a diamond, and colored red. It was not excluded, not hidden, and not labeled.

Here is a simple method to find your favorite marker:

```
Markers_dt = New Table( "Marker Codes", Add Rows( 300 ),
New Column( "Marker Number", Numeric, Formula( Row() ) ));
For( i = 1, i <= N Row( Markers_dt ), i++,
Marker Of( Row State( i ) ) = i);
```

Or, markers can be set using the **For Each Row( )** function.

```
For Each Row( Marker Of( Row State( Row() ) ) = Row() );
```

The previous script creates a data table named Marker Codes that shows available choices for markers.

This section includes some basic commands and syntax for working with row states, but it does not capture its full utility. Spend time in the *JMP Scripting Guide* studying row states, and you will be rewarded!

Save and Restore Row States

There are often times when you need to save current row states. If you want to clear row states, but retain the information for later use, it is possible to save the row states to a new column in the table. Saving row states is a two-step process. First, create a new column with the **Row State** data type, and then store the row state values. The following short script shows how to save and restore row states:

```
// Cars_dt has many row states set to preserve before clearing
rsCol = Cars_dt << New Column("Saved RowStates", Row State);
rsCol << Copy From Rowstates; //copies row states to the column
Cars_dt << Clear Rowstates;

//--Restore previous rowstates
rsCol <<  Copy to Rowstates;
```

A Few Items to Note

- Managing row states is an important JSL skill. Using row states, you can remove observations from analyses and enhance graphics with colors, symbols, and labels.
- **Exclude** and **Hide** are useful analysis statements, but toggling row states can trip up even an experienced scripter. It is easy to do the wrong thing without using some caution. Use your row state power for good, not evil!

Modify Column Information

JMP graphs and analyses generally require the variables of interest to be in data table columns. All data table columns have properties such as column name, data type, and modeling type, and optional properties such as axis settings or value ordering. In addition, numeric columns have format settings. Column properties in JSL can run the gamut, from being fun to useful to essential for your analyses. Column properties are so broad and deep that they are covered in a separate section of this chapter.

Example

Suppose that you are responsible for writing a JSL script that prepares a standard data set of failures from a manufacturing process for monthly analysis. Results are reported to a multinational team. You want to ensure that the correct columns are included, and that the data and modeling types are set appropriately. Recall that in JMP, the data type determines how values are formatted in the table, how they are stored internally, and whether they can be used in calculations. There are three data types: numeric, character, and row state. Modeling type applies only to columns that are numeric or character. The modeling type tells JMP how to use the column for analyses. There are three modeling types: continuous, nominal, and ordinal. The JMP Sample Data file Failuressize.jmp is used in this example. The script is 3_ModifyColumnInfo.jsl.

Column Names

The following script returns a list of all of the column names in the data table:

```
Failsize_dt = Open("$SAMPLE_DATA/Quality Control/Failuressize.jmp");
Colnames_lst = Failsize_dt << Get Column Names();
```

It returns the list {Process, Day, Causes, Count, IProcess} to the Log window. The list of column names is assigned the variable name Colnames_lst. This variable can be used in subsequent scripting. There is more information about variables and their names in Chapter 4, "Essentials: Variables, Formats, and Expressions," and about lists in Chapter 5, "Lists, Matrices, and Associative Arrays."

The **Get Column Names** command can also get lists of specific data and modeling types.

```
Num_lst = Failsize_dt << Get Column Names( Numeric );
Nom_lst = Failsize_dt << Get Column Names( Nominal );
```

The previous script returns {Day, Count, IProcess} as the columns with a numeric data type, and {Process, Causes} as the columns with a nominal modeling type. If you require the items in the list of numeric columns to be quoted strings rather than the column names, then use the optional **String** argument in the **Get Column Names** command.

```
Numstr_lst = Failsize_dt << Get Column Names( Numeric, String );
//returns {"Day", "Count", "IProcess"}
```

In the previous lines of code, lists of columns that are numeric and nominal are assigned to variables. This is extremely useful for analyses that use all columns in the data table with a certain data type or modeling type. For example, if you want to create a distribution for each numeric column in your data table, use the Distribution platform with a list containing all numeric columns. The **Eval( )** function is required to evaluate the argument so that it is recognized as a list of columns.

```
Failsize_dt <<  Distribution( Y( Eval( Num_lst ) ) );
```

The following script assigns a reference named Date_col to the second column. Now, actions can be taken on the assigned variable, such as changing the name from Day to Date.

```
//Rename a column
Date_col = Column(Failsize_dt, 2 );
Date_col << Set Name( "Date" );
```

Data Types

The **Get Data Type** command returns the data type of the referenced column. In the following script, it returns Numeric. The **Data Type** command sets the data type of a column. By sending the **Data Type** command to a column, and specifying one of the three data types (character, numeric, and row state), the column data type is updated. In the following script, the column Count (column 4) has its data type changed to Character, and then back to Numeric.

```
//Get and Set Data Type
Count_col = Column( Failsize_dt, 4 );
Count_col << Get Data Type;
Count_col << Data Type( "Character" );
Count_col << Data Type( "Numeric" );
```

A common error that occurs when setting a data type for a column that contains a formula is to try to change the data type to a type not supported by the formula. If formula evaluation is not suppressed, then the formula needs to be deleted before the column can be changed to a data type that is not supported by the formula. For example, suppose you want to change the column IProcess to have a character data type. The following lines of code delete the formula from the column, and then change the data type:

```
//Delete formula and set data type
IProc_col = Column( Failsize_dt, 5 );
IProc_col << Delete Formula;
IProc_col << Data Type( "Character" );
```

Modeling Types

After changing the data type of the column IProcess to character, the modeling type also changes from continuous to nominal. To confirm this change, send the **Get Modeling Type** command to the column. **Get Modeling Type** returns continuous, nominal, or ordinal. To change the modeling type of a column, send the **Set Modeling Type** command to the column.

```
//Get and Set Modeling Type
IProc_col << Get Modeling Type;
IProc_col << Set Modeling Type("Ordinal");
```

Column Formats

Column formats deserve special attention. They have their own section in Chapter 2, "Reading and Saving Data." For date formats, there is a separate section in Chapter 4. In this section, we briefly discuss column formats and how they apply to the multinational monthly report example. The **Format( )** message controls numeric, date, and datetime formatting. The first argument is a quoted string from the list of format choices in the Column Info dialog box. Additional parameters depend on your choice of format. Also, you can just set the field width.

```
Column( "Count" ) << Format( "Best", 5 );
Column( "Causes" ) << Set Fieldwidth( 30 );
```

There are many date and datetime formats available. JMP stores dates as the number of seconds since January 1, 1904. The **Format** statement determines how to read and display the date. To see the format for the date column in this example, run the following script:

```
Column( Failsize_dt, 2 ) << Get Format;
```

The code returns Format("m/d/y", 10). Because you are preparing this data set for a multinational team, you choose to display a date format that is more international. In the **Format( )** statement, the first argument is how the date is displayed, the second argument is the width, and the third argument is how to read the date.

```
Column( Failsize_dt, 2 ) << Format( "ddMonyyyy", 10, "m/d/y" );
```

Create and Delete Columns

To get the most from JSL, it is necessary to know how to create and delete columns in a data table. These commands are used frequently once a JSL user goes beyond capturing scripts written by JMP.

When creating and deleting columns in JSL, the object being changed is the data table. The data table is referenced in the script, and the send operator (<<) is used to send the changes to it. Be sure to reference the data table when creating or deleting a column. A common mistake is to send the command to the column, rather than to the data table. An object cannot delete itself, so an error is produced if the column is referenced when trying to delete it.

It is straightforward to create a new column. Use the **New Column( )** command:

```
xxxx_dt << New Column("column name", <optional arguments> )
```

There are several arguments that can be specified when creating a new column. If arguments are specified, then a name for the column must be provided. The name must be listed first, and it must be enclosed within double quotation marks. If no other arguments are specified, the default data type and modeling type are numeric and continuous, and values are missing. Table 3.2 provides some of the more commonly used column arguments, and default values.

Table 3.2 Commonly Used Column Arguments and Default Values

Arguments	Descriptions	Default
Data Type	Numeric | Character | Row State	Numeric
Modeling Type	Continuous | Ordinal | Nominal	Continuous
Width	Defines the width of a column.	Width(10)
Format	JMP formats, including date and datetime.	Best
Add Rows	Specifies the number of rows to add to the table.	No Rows
Values	Defines values or uses a list or matrix.	Missing
Formula	Fills the column with values created from a JMP formula.	None
Eval Formula	Forces JMP to evaluate the formula.	None

Column Formulas

Using a formula to populate a new column is very useful, but use caution to ensure that the formula is evaluated properly. When new columns containing formulas are created, one way to ensure that formulas are evaluated in the proper sequence is to send the **Run Formulas** command to the data table.

```
dt << Run Formulas;
```

The optional argument **Eval Formula** listed in Table 3.2 forces that column formula to be evaluated before moving to the next line of code. The section "Formulas" in Chapter 4 is devoted to formulas. It includes a discussion on the multiple methods available for controlling formula evaluation.

Example

Suppose that you work for a shoe company, and you are interested in determining the differences between two different materials used for the soles of running shoes. The JMP data table that you are given contains two poorly named columns, runner 1 and runner 2. The data are paired by runner, and the response is wear in millimeters. The data table RunningShoes.jmp is used in this example, and the script is 3_CreateDeleteColumns.jsl.

The script references the data table, renames the columns, creates new columns, and populates them using different methods (formulas and values). It also deletes columns that are no longer needed.

```
//reference data table
wear_dt = Data Table( "RunningShoes" ); //From $JSL_Companion
```

```
//reference columns and rename so the names are more meaningful
left_col = Column(wear_dt,"runner 1") << Set Name( "Wear (mm) Left
Foot" );
right_col = Column(wear_dt,"runner 2") << Set Name( "Wear (mm) Right
Foot" );

/*define new columns: change units to microns and add runner ID */
left_mu_col = wear_dt << New Column( "Wear (microns) Left Foot",
   Numeric, Continuous, Format( "Fixed Dec", 10, 0 ),
   Formula( :Name( "Wear (mm) Left Foot" ) * 1000 ));

right_mu_col = wear_dt << New Column( "Wear (microns) Right Foot",
   Numeric, Continuous, Format( "Fixed Dec", 10, 0 ),
   Formula( :Name( "Wear (mm) Right Foot" ) * 1000 ));

diff_mu_col = wear_dt << New Column( "Wear Diff (microns)",
   Numeric, Continuous, Format( "Fixed Dec", 10, 0 ),
   Formula( :Name( "Wear (microns) Left Foot" )
       - :Name( "Wear (microns) Right Foot" ) ));

wear_dt << New Column( "Runner", Numeric, Ordinal,
   Values( 1 :: N Rows() ) );

//ensure formulas are run
wear_dt << Run Formulas;

/* delete wear (mm) columns - the columns to be deleted contain
formulas so the formulas need to be deleted before the columns can be
deleted. */
right_mu_col << Delete Formula;
left_mu_col << Delete Formula;
left_col << Set Selected( 1 );
right_col << Set Selected( 1 );
wear_dt << Delete Columns;
```

A Few Items to Note

- There are multiple ways to populate values for a new column. This script uses formulas and values. The commands **Set Each Value( )** and **For Each Row( )** are alternate commands for populating values.
- If the formulas are evaluated and not suppressed, then, when you want to delete a column that is used in a formula, the formula must be deleted first.
- Column properties, such as name and data type, are metadata that belong to the column. This information is sent to the column. Creating and deleting columns are commands that act on the data table, and they need to be sent to the data table. Tables also have metadata, such as name and table properties.
- The **Get Script** command is useful for JMP users who are learning JSL. Create a column with properties using the point-and-click menus. Using the send **(<<)** operator, submit the **Get Script** command to that column. The JSL syntax is written to the Log window. To get

the syntax for the column containing wear of the left shoe in microns, run the following line of code:

```
left_mu_col << get script;
```

Modify and Subset Portions of Tables

This section is partitioned into three subtopics: subset command syntax; get, set, and clear column and row selections; and assign values to selected rows.

Subset Command Syntax

The **Subset( )** command creates a new data table from a source table. Figure 3.2 displays the JMP dialog box for **Tables ► Subset** for the current data table. For this example, the Sample Data table is $Sample_Data\Candy Bars.jmp, and example scripts for this subtopic are in 3_SubsetOptions.jsl. Syntax for the **Subset( )** command includes options for features in the dialog box. In addition, the **Subset( )** command syntax allows you to specify the source table. If the script does not send the **Subset( )** command to a specific table, then the current data table is used as the source table. Compare the contents of the dialog box to the **Subset( )** command syntax, as documented in the *JMP Scripting Guide*.

Figure 3.2 Subset Dialog Box: (L) Default (R) Stratified Sample, 2 for Each Brand

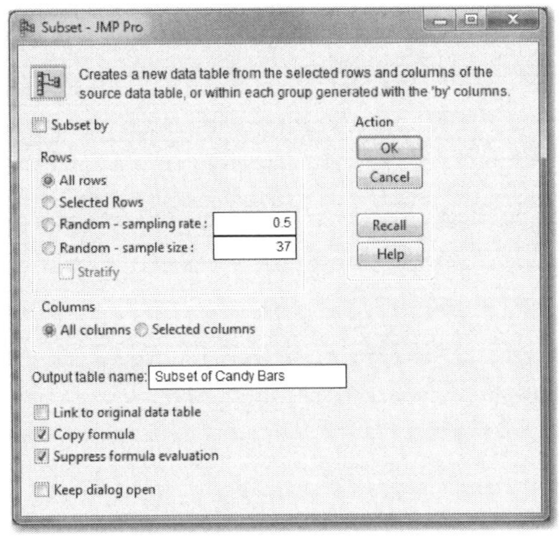

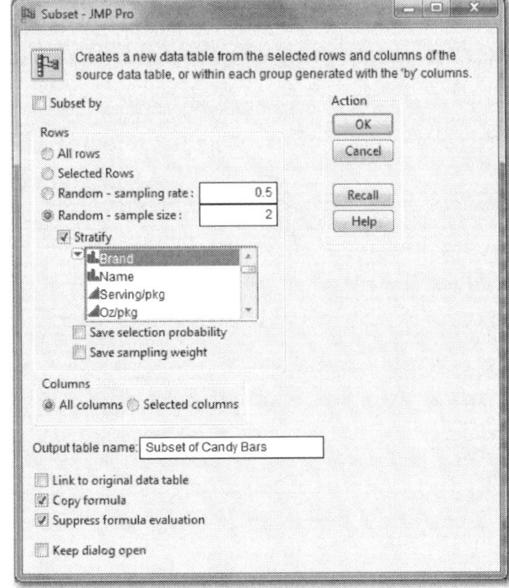

Here is the general syntax for the **Subset( )** command:

```
subdt = dt << Subset(
  Columns( column list ),     /*[1]*/
  Rows( row matrix ),         /*[2]*/
  Sampling rate( n ),         /*[3]*/
  Output Table Name( "name" ),
  Linked,                     /*[4]*/
  Copy Formula( 1 | 0 ),
  Suppress Formula Evaluation( 1 | 0 )
);
```

[1] **Columns(** column list **)** If this line is eliminated, or if a column list is not specified, then only columns currently selected in the source table, dt, are written to the new table. If no columns are selected in dt, all columns are extracted. As of JMP 9, a vector of column numbers can be used. JMP 8 requires a list. For example, JMP 9 syntax: Columns([2, 3, 7]). JMP 8 syntax: Columns(As List ([2, 3, 7])).

[2] **Rows(** row matrix **)** If rows are not specified and a sampling rate is not specified, then all rows currently selected in the source table, dt, are extracted. If no rows are selected in dt, all rows are extracted.

[3] **Sampling rate(** n **)** accommodates two options from the dialog box. If 0 < n < 1, then a proportion of the table is extracted. If n is a positive integer, then n rows are extracted. The **Subset( )** command also allows the option **Sample Size(** n **)**. We recommend using **Sampling rate** when a proportion (0 < n < 1) of the table is wanted, and **Sample Size** when you specify the number of rows. If **Sample Size(** n **)** is used, and n equals the number of rows in the table, the returned table is a shuffle of the original table.

Sampling rate(0 **)** produces an empty result. As the buttons imply, only one row option is applied, and it is either sampling rate or sample size. If your script's **Subset( )** command includes both of these options, then the option that appears first is applied.

[4] **Linked** is not a Boolean option. Eliminate line [4] for a subset table that is not linked to the source.

Figure 3.2 depicts other options in the dialog box.

Keep Dialog Open has very little value when scripting because the dialog box is not open to begin with.

All Rows and **All Columns** are not options currently available in JMP 9. However, specifying **Selected columns(** 0 **)** ignores column selections and extracts all columns. This functionality is undocumented, and has the same effect as the dialog box button labeled **All Columns**. Specifying

Columns(1 :: N Col (dt) **)** also returns all columns. 1 :: N is equivalent to the **Index(** 1, N **)** function that returns the matrix [1, 2, 3,…, N].

Specifying **Selected Rows(** 0 **)** has no effect in JMP 9. Using **Sample Size(** N Row (dt) **)** or **Sampling rate(** 1 **)** ignores current row selections. However, the resulting table is shuffled (see [3] above). To return all rows with no shuffling, specify **Rows(1 :: N Row** (dt)).

Subset by is a check box that, when selected, displays a column list. One or more columns can be selected, and a new table is created for each unique combination of the values of the selected columns. To deploy this functionality in your script, specify **By(**col 1, col 2, …**)**. An example is provided in Chapter 8, "Custom Displays," when discussing how to build a multivariable display.

Stratify is a check box that is selected in Figure 3.2 (R). In addition, column **Brand** is selected. The following script shows this option. The **Stratify** feature is used with random sample options.

If either **Sample Size(** n **)** or **Sampling Rate(** r **)** is used as an option, then **Stratify(** Col1, Col2, …**)** can be used. The **Sample Size** option enables two additional options to save columns to the resulting subset table. These options are the selection probability or the sampling weight. The argument for each of these options is used to name the saved column.

```
Data Table( "Candy Bars" ) << Subset(
  Copy formula( 0 ),
  Sample Size( 2 ), Stratify( :Brand ),
  Save selection probability( Selection Probability ),
  Save sampling weight( Sampling Weight ));
```

This script produces a table with the following characteristics:

- It is named Subset of Candy Bars because no **Output Table Name( )** was provided.
- It contains all 19 columns (if none were selected).
- It is not linked to the source table Candy Bars.
- It copies values, but not formulas if any exist.
- It contains 26 of the 75 rows from the source table. The Brand column from the source table Candy Bars contains 18 different brands, 10 of which appear just once. That is, 10 brands list 1 candy bar, and 8 brands list 2 or more candy bars. A stratified sample with a sample size of 2 randomly selects 2 from each of the 8 product brands, and 1 from the others [26 = 2*8 + 10]. Figure 3.3 shows the two new columns, Selection Probability and Sampling Weight, created in the subset table. The Hershey brand has 29 candy bars. With a sample size of 2, each candy bar has a selection probability of 2/29=0.068965. Its sampling weight is the reciprocal value, 29/2=14.5.

Figure 3.3 Selection Probability and Sampling Weight for Stratified Sample

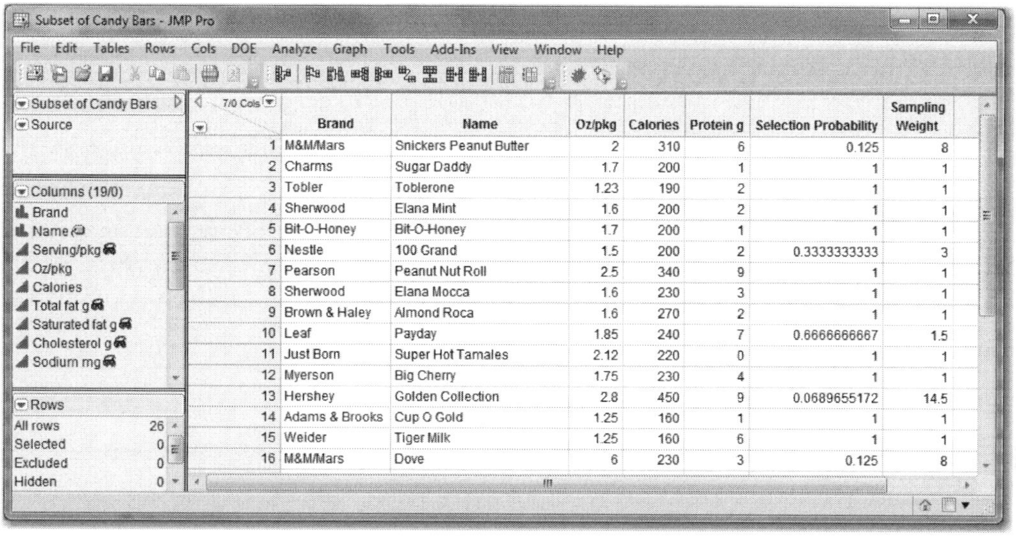

	Brand	Name	Oz/pkg	Calories	Protein g	Selection Probability	Sampling Weight
1	M&M/Mars	Snickers Peanut Butter	2	310	6	0.125	8
2	Charms	Sugar Daddy	1.7	200	1	1	1
3	Tobler	Toblerone	1.23	190	2	1	1
4	Sherwood	Elana Mint	1.6	200	2	1	1
5	Bit-O-Honey	Bit-O-Honey	1.7	200	1	1	1
6	Nestle	100 Grand	1.5	200	2	0.3333333333	3
7	Pearson	Peanut Nut Roll	2.5	340	9	1	1
8	Sherwood	Elana Mocca	1.6	230	3	1	1
9	Brown & Haley	Almond Roca	1.6	270	2	1	1
10	Leaf	Payday	1.85	240	7	0.6666666667	1.5
11	Just Born	Super Hot Tamales	2.12	220	0	1	1
12	Myerson	Big Cherry	1.75	230	4	1	1
13	Hershey	Golden Collection	2.8	450	9	0.0689655172	14.5
14	Adams & Brooks	Cup O Gold	1.25	160	1	1	1
15	Weider	Tiger Milk	1.25	160	6	1	1
16	M&M/Mars	Dove	6	230	3	0.125	8

When scripting, and especially when using the **Subset( )** command, managing table selections is very important. To ignore table row and column selections, the scripter needs to specify columns and rows to be extracted. The 3_SubsetOptions.jsl script provides examples of items [1]—[4] and each option described in this section.

Get, Set, and Clear Column and Row Selections

In JSL, columns and rows are different objects, and the commands to do similar tasks are often different. For example, earlier in this chapter, you learned the commands for selecting rows. You know that multiple rows can be selected with one command. However, columns must be selected one at a time.

The following snippets from the 3_GetSetClearSelections.jsl script demonstrate how to select all of the columns in the $Sample_Data/Candy Bars.jmp data table that have grams as units. They also select all rows (candy bars) where protein per ounce is less than or equal to 2. This last selection is based on a calculation—the ratio of two column values. The **Get** commands return lists of selections. The **Clear** commands deselect the selections.

```
Candy_dt = Open( "$SAMPLE_DATA\Candy Bars.jmp" );
//--get all names as text
all_Names = Candy_dt << Get Column Names( string );
```

```
//select all columns with grams as units, i.e., name ends with "g"
//--use set selected
For( i = 1, i <= N Items( all_Names ), i++,
   If( Ends With( all_Names[i], " g" ),
   Column( i ) << Set Selected( 1 ))
);

//--select rows(candy bars) with <= 2 g protein/oz, 49 of 75 rows
Candy_dt << Select Where( :Protein g / :Name( "Oz/pkg" ) <= 2  );

//--get a list of selected columns and a list of selected rows
usr_SelectedCols = Candy_dt << Get Selected Columns;
usr_SelectedRows = Candy_dt << Get Selected Rows;

//----clear row selections
Candy_dt << Clear Select;
//clear column selection ...note no trailing S on selection or column
Candy_dt << Clear Column Selection;

//--reset previously selected rows
Candy_dt << Select Rows( usr_SelectedRows );

//--reset previously selected columns
For( i = 1, i <= N Items( usr_SelectedCols ), i++,
   usr_SelectedCols[i] << Set Selected( 1 )
);
```

Current table selections can be maintained by using **Get Rows Where( )** instead of **Select Where( )**, and then listing specific rows and columns when creating a table subset. Another option is to follow these steps:

1. Create lists of selected rows and selected columns.
2. Clear user selections.
3. Do what needs to be done.
4. Reset the original selections.

Maintaining user settings is a hallmark of a professional program. The 3_GetSetClearSelections.jsl script includes an example using these steps and alternatives. This script also demonstrates the syntax for using column references in logical expressions and formulas. This syntax is the variable name, followed by empty brackets (for example, protCol[ ]). Using variable references is discussed in Chapter 4. An example is provided in this section as a preview of its use for selecting rows.

```
//Suppose you are scripting with references to columns
protCol = Column( Candy_dt, "Protein g");
ozCol = Column( Candy_dt, "Oz/pkg");
//----clear row selections
Candy_dt << Clear Select;

// convert this command:
// Candy_dt << Select Where( :Protein g / :Name( "Oz/pkg" ) <= 2  );
Candy_dt << Select Where( protCol[] / ozCol[] <= 2  );
```

Assign Values to Selected Rows

Now that you understand the commands for getting rows conditionally, the next step is to learn how to assign values. Perform this task. Open and reference the Candy Bars.jmp data set, create a new column for labels, identify candy bars that have more than 2 grams of protein per ounce, and give those candy bars a label of Good Protein.

The following script from 3_AssigningValues.jsl documents two methods for selecting rows. These two methods are **Select Where( )** with **Get Selected Rows( )** and **Get Rows Where( )**. Both methods return the same result, which is a matrix of row numbers that satisfy the condition. Recall that **Get Rows Where** has the benefit of not changing the current highlighting and selections in the table.

```
//--Open and reference data and create a new column
//$SAMPLE_DATA/ posix format -- see script 2_Paths.jsl
Candy_dt = Open( "$SAMPLE_DATA/Candy Bars.jmp" );
MyPicks_col = Candy_dt << New Column( "MyOpinion", character, nominal
);

//--Select rows in the table and get selected rows
//Uses a formula to select
Candy_dt << Select Where( :Protein g / :Name( "Oz/pkg" ) > 2 );
RowChoice = Candy_dt << Get Selected Rows;
Candy_dt << Clear Select;   //Clear row selections

//--Get selected rows *without* changing current selections
RowChoice2 = Candy_dt << Get Rows Where(:Protein g/:Name( "Oz/pkg") >
2 );

//----Assign a constant value to chosen rows in new column MyPicks
::MyPicks_col[ RowChoice2 ] = "Good Protein" ;

//Name Selection in Column is a new funtion in JMP 9
//It can create this new column in 1 command
Candy_dt << Select Rows( RowChoice ) ;
Candy_dt << Name Selection in Column(
      Column Name( "New Choice" ),
      Selected( "Good Protein" ),
      unSelected() //or specify a value for unselected rows
    );
//----Graph Calories vs. Carbohydrates using MyOpinion as a legend
```

Figure 3.4 Bivariate Graph with Assigned Values Legend

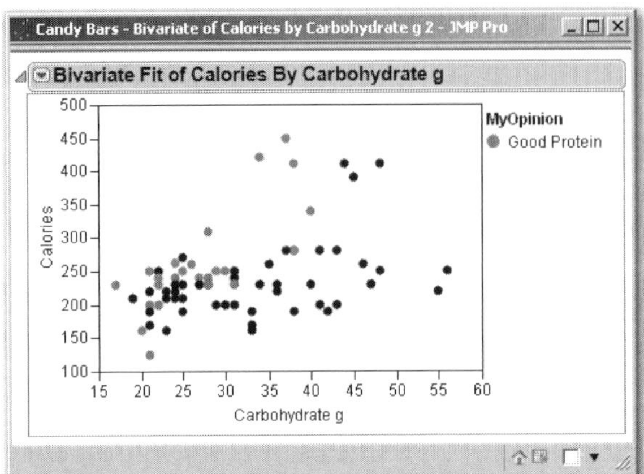

Previous sections in this chapter provided examples for assigning values to an entire column using column messages **Set Values( )**, **For Each Row( )**, and **Set Each Value( )**, and using formulas. So far, this section has demonstrated how to assign a constant value for selected rows in a column using the following syntax:

```
colRef[ rowlist ] = value;
```

There is no comparable command for rows. Unless you are assigning all values in a column, or you are assigning a selected set of rows to a single value, values are assigned by cell—a specific row and column. Suppose there is a need to update the values of the Candy Bars data set: M&M/Mars Milky Way Lite has a new, lighter recipe. And, Pearson Peanut Nut Roll has reduced carbohydrates and sodium. More than likely, the easiest method to manage these updates is **Tables ▶ Update**. However, if different columns are being updated for each row, it can be tricky. The following syntax offers one option. The 3_AssigningValues.jsl script includes the following code and an alternate example. Remember, code is written for the task. For updating specific cells in a table, the best method depends on the format of the incoming updates.

```
//-------------Update information for two candy bars----------------
selRow = Candy_dt << Get Rows Where(:Name == "Milky Way Lite");
Row() = selRow[1];  //set the current row to be the found row
Eval List( {:Calories = 150, :Total fat g = 4, :Saturated fat g =
3});

selRow = Candy_dt << Get Rows Where(:Name == "Peanut Nut Roll" &
   :Brand == "Pearson");
Row() = selRow[1];  //set the current row to be the found row
Eval List( {:Calories = 320, :Sodium mg = 200, :Carbohydrate g = 38,
   :Sugars g = 27});
```

Restructure Tables

One of the most powerful features of JMP is the ability to easily restructure and manipulate data tables. Capturing the JSL code to do this is simple. But, this feature is often overlooked by users who are just getting familiar with scripting. For most of the commands in the **Tables** menu, a new table is generated when these commands are executed. (The exceptions are when you select **Sort** and replace the table or **Tabulate**.) As shown in Figure 3.5, when a new table is generated, the JSL command can be captured by looking at the table property **Source** in the new JMP table.

Figure 3.5 Table Summary with Source Table Property in the Table Panel

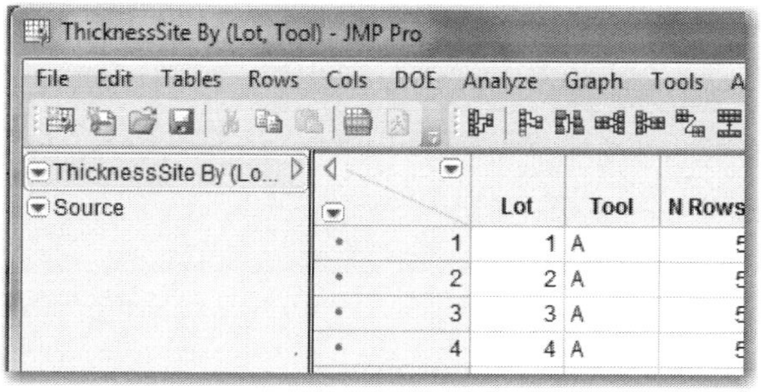

ThicknessSite By (Lot, Tool) - JMP Pro

File Edit Tables Rows Cols DOE Analyze Graph Tools A

ThicknessSite By (Lo...

Source

	Lot	Tool	N Rows
1	1	A	5
2	2	A	5
3	3	A	5
4	4	A	5

Example

Consider the task of writing a script to analyze thickness data from a semiconductor manufacturing operation. The goal is to determine whether the spatial pattern across the wafer is similar for the two tools. In addition, you want to compare tool stability, the lot level means, and standard deviations. There are 20 lots from each tool, and five sites are measured on each wafer. The ThicknessSite.jmp data table contains columns for Tool, Lot, Site, and Thickness. It is in a stacked format, meaning that each variable is in a single column. A second data file, SiteCoordinates.jmp, contains information that translates the site location names (left, center, etc.) into wafer site coordinates (x, y) relative to the wafer's center. The script containing all of the steps in the analysis is named 3_RestructureTables.jsl.

Begin with a variability graph because the original data table is in a stacked format. A variability graph is created with Thickness as the response, and Tool and Site as the grouping variables.

To compare the tools at the lot level, the data must be summarized. (In the 3_RestructureTables.jsl script, the original data table is referenced as thick_dt.)

```
//Summarize the data to the lot level
thick_sum_dt = thick_dt << Summary(
Group( :Lot, :Tool ), Mean( :Thickness ), Std Dev( :Thickness ));
```

The **Summary** command generates a new data table that stores the summary statistics.

Using this summarized data table, a one-way analysis is performed on the lot level means and standard deviations. The next step is to generate additional analyses that require the data to be in a split or unstacked format. **Split** maps several rows in one column to one row in several columns.

```
//Split the table in order to do additional analyses
thick_split_dt = thick_dt << Split(
Split By( :Site ), Split( :Thickness ),
Group( :Lot, :Tool ), Output Table Name( "ThicknessSite Split" ));
```

The **Split** command generates a new data table that stores the five rows for each Lot and Tool as one row with thickness values in five columns named Bottom, Top, Right, Left, and Center.

When splitting a data table, it is critical that the **Group** variables are defined appropriately. These grouping variables ensure that the row integrity of the data table is maintained. In this example, maintaining row integrity means that the thickness values for the five sites are associated with the correct lot and tool. For a large table with many variables, determining the correct grouping variables is not always intuitive. It is helpful to do a quick check of a newly split table to be certain that the split was done correctly, and that row integrity was maintained.

Once the table is split, an overlay plot is drawn. It shows thickness versus lot number with sites overlaid, and a vertical line drawn with the tool labeled. (See Figure 3.6.) Correlations are performed using the unstacked data.

Figure 3.6 Overlay Plot of Site Thickness versus Lot and Tool Using **Tables ▶ Split**

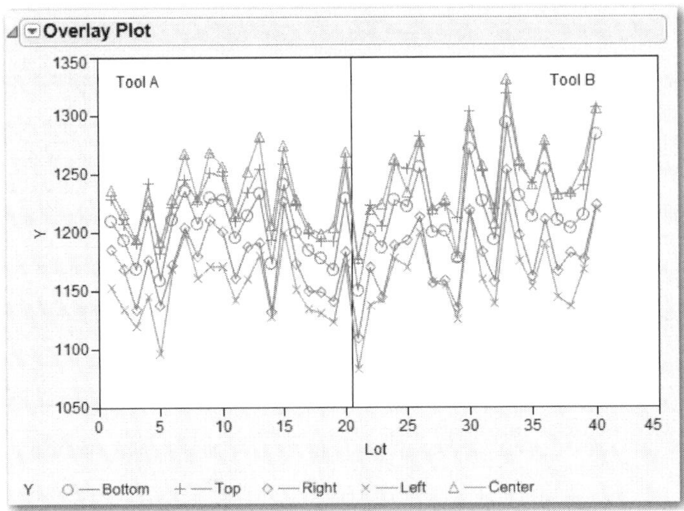

The final step in the analysis is to create contour plots to assess the pattern across the wafer. Because site locations on the wafer are needed, the original data table is modified to include the wafer-level coordinates for the sites labeled Bottom, Top, Right, Left, and Center. This is performed using the second data table, SiteCoordinates.jmp, and the **Update** command.

```
//Update current table to include x and y coordinates
thick_dt << Update( With( Data Table( "SiteCoordinates" ) ), Match
Columns( :Site = :Site ) );
```

Rather than using the **Update** command, the **Join** command can be used to augment the original data table with the coordinates. The **Join** command has more options, and it is more flexible than the **Update** command. The syntax for the **Join** command is provided in the 3_RestructureTables.jsl script. Joining tables is a common analysis task, and there are multiple methods (one to one, many to many, etc.). The next section covers the **Join** command in detail.

A Few Items to Note

- A single analysis session can generate numerous tables. To avoid unintended consequences, it is important to explicitly send messages (using the send **(<<)** operator) to the table being modified.
- When a data table is created, it is helpful to provide it with a variable name. Typically, you will need to reference a newly created table later in the script, and the variable name will be needed. If **Output Table Name** is not specified, JMP provides a name—Untitled n. In the case of the **Summary** command, the new table name is a concatenation of the original table name with **Group** by column names. This concatenated table name becomes unwieldy with multi-variable, multi-pass summaries and long variable names.
- **Summary** is a table command (message) that creates a new table of aggregated statistics. **Summarize( )** is a JSL function that can also be used to compute aggregated statistics. However, the results are returned as global variables, rather than in a data table. The **Summarize** function is covered in later chapters.

Join Tables

The previous section focused on JMP capabilities for restructuring tables. It briefly discussed joining tables using the **Update** command. Because joining tables is a frequently used and powerful feature, the **Join** command deserves its own section. There are several methods to join tables, and each method has its benefits and practical uses. The general structure for joining tables in JSL follows. **Select( )**, **Select With( )**, **Drop multiples( )**, and **Include non-matches( )** are optional Join arguments.

```
first_dt << Join( With( second_dt ), < Select() >, < Select With() >,
By join type, < Drop multiples() >, < Include non-matches() >);
```

The join types are Row Number, Cartesian Join, and Matching Columns. Selecting the columns to include from the original table is also possible with the options **Select** and **Select With**. The options available for joining tables in JSL align well with standard SQL equijoins.

Table 3.3 Types of Joins

SQL Join Type	JSL Join Type	JSL Syntax for Include non-matches
Inner Join	Matching Columns	(0, 0)
Cross	Cartesian Join	NA
Left Outer	Matching Columns	(1, 0)
Right Outer	Matching Columns	(0, 1)
Full Outer	Matching Columns	(1, 1)

The following two tables, Big Class Joins Table1.jmp and Grade Level Joins Table2.jmp, are used in subsequent examples to show the differences in join types. The script is included in 3_JoiningTables.jsl. There is a missing value in Figure 3.7, and ROBERT is in Figure 3.8, but not in Figure 3.7.

Figure 3.7 Joins Table 1

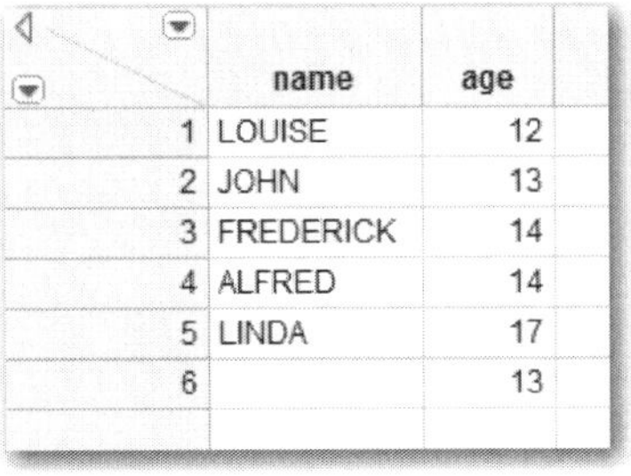

	name	age
1	LOUISE	12
2	JOHN	13
3	FREDERICK	14
4	ALFRED	14
5	LINDA	17
6		13

Figure 3.8 Joins Table 2

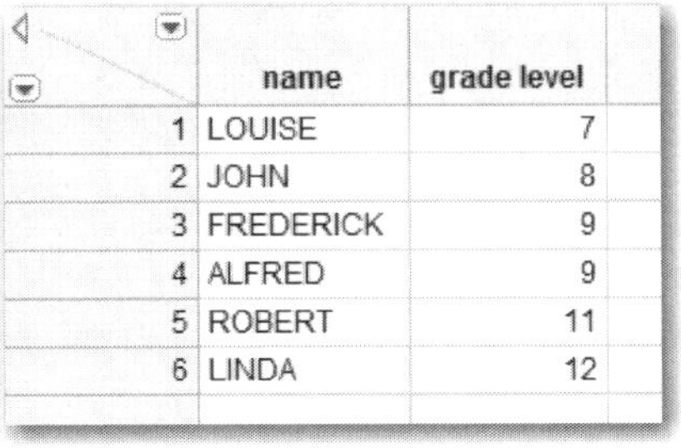

	name	grade level
1	LOUISE	7
2	JOHN	8
3	FREDERICK	9
4	ALFRED	9
5	ROBERT	11
6	LINDA	12

Inner Join

An inner join is the most commonly used join type, and it is often the default join in applications. To perform an inner join in JSL, use the **By Matching Columns( )** option, and **(0, 0)** for the **Include non-matches** option.

```
Data Table( "Big Class Joins Table1" ) << Join(
With( Data Table( "Grade Level Joins Table2" ) ),
By Matching Columns( :name = :name ), Drop multiples( 0, 0 ),
Name( "Include non-matches" )(0, 0),Output Table Name("Inner Join"));
```

Figure 3.9 Inner Join

	name of Big Class Joins Table1	age	name of Grade Level Joins Table2	grade level
1	ALFRED	14	ALFRED	9
2	FREDERICK	14	FREDERICK	9
3	JOHN	13	JOHN	8
4	LINDA	17	LINDA	12
5	LOUISE	12	LOUISE	7

Figure 3.9 is the result of a basic inner join. The missing value that was in Figure 3.7 is not included, and ROBERT, who did not have a match in Figure 3.7, is dropped. If you want to drop the name column that is included twice in the joined table, then use an option for selecting columns. This type of join is often referred to as a natural join.

```
Data Table( "Big Class Joins Table1" ) << Join(
With( Data Table( "Grade Level Joins Table2" ) ),
Select( :name, :age ), SelectWith( :grade level ),
By Matching Columns( :name = :name ),
Drop multiples( 0, 0 ),
Name( "Include non-matches" )(0, 0));
```

Cartesian Join

A Cartesian, cross, or product join takes all of the elements in the first table, and combines them with all of the elements in the second table. It results in a table with all possible combinations of the elements.

Figure 3.10 shows only a portion of the resulting table. The table has 36 rows.

```
Data Table( "Big Class Joins Table1" ) << Join(
With( Data Table( "Grade Level Joins Table2" ) ),
Cartesian Join, Output Table Name("Cartesian Join"));
```

Figure 3.10 Cartesian Join

	name of Big Class Joins Table1	age	name of Grade Level Joins Table2	grade level
1	LOUISE	12	LOUISE	7
2	LOUISE	12	JOHN	8
3	LOUISE	12	FREDERICK	9
4	LOUISE	12	ALFRED	9
5	LOUISE	12	ROBERT	11
6	LOUISE	12	LINDA	12
7	JOHN	13	LOUISE	7
8	JOHN	13	JOHN	8

Outer Joins

Outer joins do not require that the two joined tables have matching values. The joined table retains each value, even if a matching value is not found. By using the binary options for the **Include non-matches** option, specific outer joins are defined. The option **Include non-matches(1, 0)** retains all of the values from the first table, and drops unmatched values from the second table. This is commonly known as a left outer join. The option **Include non-matches(0, 1)** is a right outer join, and retains all of the values from the second table. A full outer join retains all of the values from both tables. In Figure 3.11, a left outer join is illustrated.

Figure 3.11 Left Outer Join

	name of Big Class Joins Table1	age	name of Grade Level Joins Table2	grade level
1		13		•
2	ALFRED	14	ALFRED	9
3	FREDERICK	14	FREDERICK	9
4	JOHN	13	JOHN	8
5	LINDA	17	LINDA	12
6	LOUISE	12	LOUISE	7

Left Outer Join

```
Data Table("Big Class Joins Table1") << Join(
With(Data Table("Grade Level Joins Table2")),
By Matching Columns( :name = :name ),
Drop multiples( 0, 0 ),
Name( "Include non-matches" )(1, 0),
Output Table Name("Left Outer Join"));
```

Duplicate Records

Duplicate measurements can be removed from a table when a join is performed and the **Drop Multiples** option is specified. A practical application of this option is pulling only the most recent record from a table. When a table is joined with itself using **By Matching Columns**, and then specifying **Drop multiples** for both tables, it retains the unique records from the table. The matching columns identify unique observations, and the unique observations found first in the data table are retained. The table should be sorted so that the record of interest appears first. (For example, to keep the most recent record, sort by descending date so that it is listed first in the table.)

Example

The objective of this example is to get a subset of the data table. The subset contains only the most recent measurements from a batch of production material called a lot. The table is sorted in descending order by the Start_Date, and then it is joined with itself. This example uses the

NoDup.jmp data file. Figures 3.12 and 3.13 show the original data table and the table after the duplicate observations are removed. For large data tables, this method is not efficient.

Figure 3.12 Table with Duplicate Values

	Lot	Start_Date	Value
1	2	06/18/2008 6:59:14 AM	998.973
2	2	06/17/2008 11:21:26 AM	1062.797
3	4	06/22/2008 2:15:36 AM	1035.431
4	3	06/19/2008 3:44:04 PM	1011.934
5	1	06/11/2008 3:35:57 AM	1009.451
6	1	06/11/2008 11:55:04 AM	1012.336
7	4	06/22/2008 5:17:41 AM	1030.479
8	1	06/11/2008 7:27:24 PM	1017.954
9	3	06/19/2008 1:59:50 PM	999.108

```
sort_dt = Data Table("NoDup")
   << Sort( By( :Start_Date ),Order( Descending ) );
sort_dt << Join( With( sort_dt ), By Matching Columns( :Lot = :Lot ),
   Drop multiples( 1, 1 ), Name( "Include non-matches" )(0, 0));
```

Figure 3.13 Table without Duplicate Values

	Lot	Start_Date	Value
1	1	06/11/2008 7:27:24 PM	1017.954
2	2	06/18/2008 6:59:14 AM	998.973
3	3	06/19/2008 3:44:04 PM	1011.934
4	4	06/22/2008 5:17:41 AM	1030.479

Reference and Apply Column Properties

Columns have assignable properties such as formulas, notes, axis values, range checking, and many more. Table 3.4 at the end of this section contains brief descriptions of the column properties available in JMP. The JMP help book *Using JMP* has detailed and complete descriptions of column properties. The JSL commands for working with column properties match the point-and-click options available in the **Column Properties** menu. (See Figure 3.14.)

Figure 3.14 Commands in the Column Properties Menu

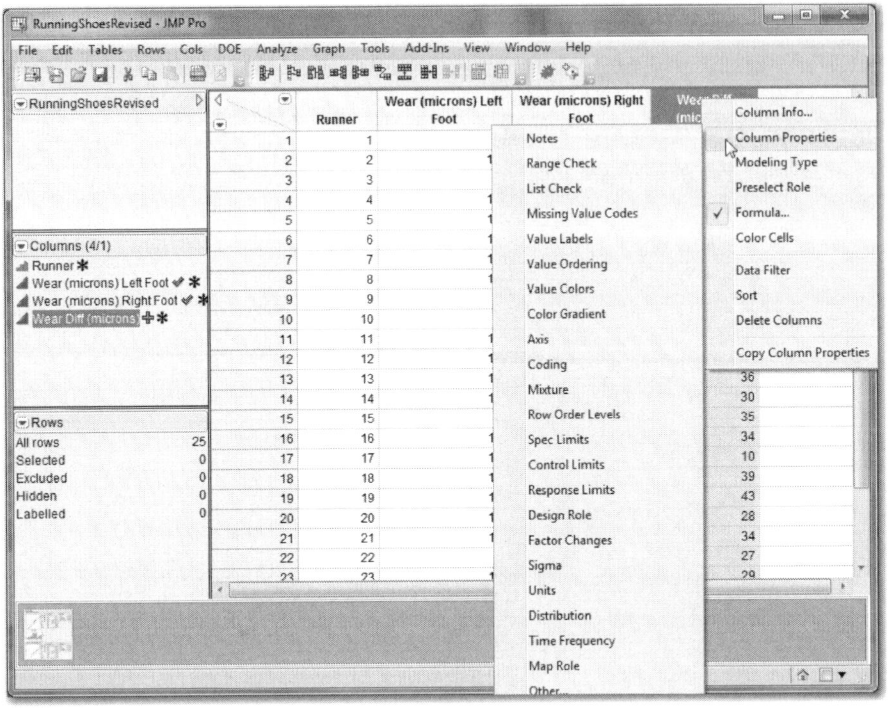

Example

This example is a continuation of an example earlier in this chapter. It references the new table that is created after running the script from that section. For this example, column properties are set to help determine whether there is a difference in wear between the two materials that were randomly assigned to each foot of the runner. The RunningShoesRevised.jmp data table is used in this example, and the script is 3_ColumnProperties.jsl.

The following code references the data table, performs a range check on the Wear columns, assigns a warning note about missing values, defines minimum and maximum axis values for the column showing differences, and creates distribution plots of all continuous variables in the data table.

```
//reference data table
wear_rev_dt = Data Table( "RunningShoesRevised" );
Current Data Table(wear_rev_dt);

//Set Range Check property on Wear columns
//Alternative Syntax: left_col << Set Property("Range Check",
LTLE(0,250))
left_col = Column("Wear (microns) Left Foot") << Range
Check(LTLE(0,250));
```

```
right_col=Column("Wear (microns) Right Foot")<<Range
Check(LTLE(0,250));

left_col << Set Property(
   "Note on Range Check",
   Values outside the range check are set to missing.
   See the log window to determine if points are outside the range.);

right_col << Set Property("Note on Range Check",
   Values outside the range check are set to missing.
   See the log window to determine if points are outside the range.);

//reorder columns so that they are sorted by name
wear_rev_dt << Reorder by Name();

//define maximum for axis of difference
diff_col = Column( wear_rev_dt, "Wear Diff (microns)" );
max_axis_diff=Ceiling(ColMaximum(diff_col)+0.5*Col Std
Dev(diff_col));
min_axis_diff=Floor(ColMinimum(diff_col)-0.5*Col Std Dev(diff_col));

//set column max axis property on difference
diff_col << Set Property("Axis",
{Min( min_axis_diff ), Max( max_axis_diff ), Inc( 5 ), Minor Ticks( 0
), Show Major Ticks( 1 ),
Show Minor Ticks( 1 ), Show Major Grid( 0 ), Show Minor Grid( 0 ),
Show Labels( 1 ), Scale( Linear )});

//select continuous columns and create distribution plots
cont_list = wear_rev_dt << Get Column Names( Continuous );
Distribution(Stack( 1 ),Column(Eval(cont_list)),Horizontal Layout( 1
),
Vertical( 0 ), Count Axis( 1 ), Normal Quantile Plot( 1 ));

//Add comment on how shoes were assigned to runners
Column( "Runner" ) << Set Property("Runner Comments",
Runners were randomly selected from a population and shoes were
randomly assigned to each foot.);

//Get comment from Runner column and print to log
runner_comments = Column("Runner")<<Get Property( "Runner Comments"
);
Print( runner_comments );
```

A Few Items to Note

- The **Range Check( )** column property cannot be used on a column that includes a formula.
- If points are outside the range when **Range Check( )** is set, the values are set to missing. A note is sent to the Log window.
- When setting axis values or other column properties, variables can be used for the values.
- Custom column properties are available. When scripting, make sure that the correct keywords are used when setting column properties. If the wrong keyword or phrase is used, then an unintended custom column property is created without an error.

- Scripting analyses on columns of a specific modeling type is easy. Just create a list of all columns of that type. Lists, which were introduced in Chapter 1, "Getting Started with JSL," are very powerful in JSL. They are discussed in more detail in Chapter 5.
- The **Other** column property is used for setting custom column properties. A script named 3_ColumnPropertyOther.jsl is provided that uses this property to define the reference mean and standard deviation for the continuous columns in a data table. The script uses a modified version of the Big Class.jmp Sample Data table. The modified version is called BigclassCustomProperties.jmp. The custom column properties are then referenced when running the Distribution platform to obtain a one-sample test on the reference mean and standard deviation. A JSL script can be included in the **Other** column property, and then run from another script.
- If a **Notes** column property is titled Notes, remember that the message included in this property is visible when opening the data table.

Table 3.4 Column Properties

Column Property	Brief Description
Formula	Adds a JMP formula to a column in a data table.
Notes	Adds comments to a column in a data table.
Range Check	Ensures that values in a column are within certain constraints.
List Check	Ensures that values in a column are in a provided list.
Value Labels	Displays a descriptive label for an abbreviated class.
Value Ordering	Defines the order in which values appear when sorted.
Value Colors	Defines the color codes for graphs using nominal and ordinal data.
Axis	Defines axis properties (Min, Max, Scale, Increment, etc.).
Coding	Defines high and low values, and codes them +/- 1 to make parameter estimates more meaningful when fitting a model.
Mixture	Specifies that the column is part of a mixture in which it and the other columns are summed to a given value.
Row Order Levels	Sorts levels of the column by the order of occurrence in the data rather than alphabetically.
Spec Limits	Triggers a capability analysis if the Distribution platform is used.
Control Limits	Defines control limits for a selected control chart type.
Response Limits	Defines the acceptable range for a response and appropriate options.
Design Role	Defines the role of a variable when the column is used to fit a model.
Factor Changes	Defines whether a factor in a design is easy, hard, or very hard to change.
Sigma	Is used by applications, such as the Control Chart platform, that require a sigma value for computations. If no value is specified, the value is calculated from the data.

Column Property	Brief Description
Units	Specifies the measurement unit for the column.
Distribution	Sets the distribution type that is fitted for the column.
Time Frequency	Defines the interval in which data is reported in the Time Series platform.
Other...	Is used to create a custom column property. You can define a name and add a JSL script as a column property.

Data Table Variables and Properties

JMP data tables store information about the data table by using table variables and properties. This type of data on data is commonly referred to as *metadata*. In the *JMP Scripting Guide*, information about table variables and properties is found by looking in the index for metadata. Table variables store text strings or numeric constants. Two common uses of table variables are adding notes about the data (such as the last time it was updated) or referencing the origin of the data, and defining variables that are used in a column formula.

Table properties are similar to table variables in that they are stored in the table panel and are readily available to the user. (See Figure 3.15.) Table properties can store JSL scripts. In JMP 9, table properties are called table scripts. From a report, when you select **Save Script to Data Table**, the script that generated the report is saved to the data table as a table property. The script used in this example is 3_DataTableProperties.jsl.

Figure 3.15 Table Panel with Table Variables and a Script

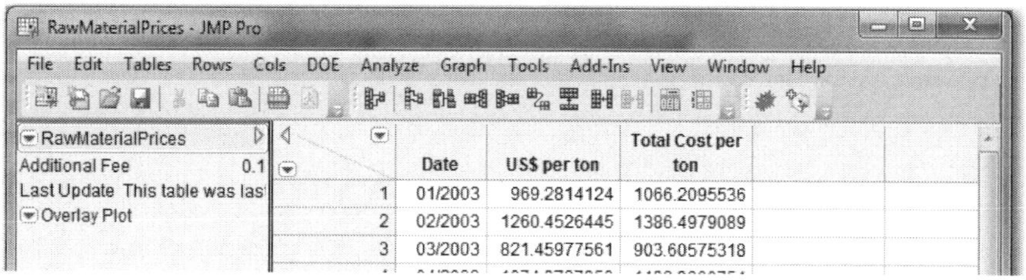

JSL can access all of the information contained in table variables and properties. For example, table variables and properties can be added, deleted, or modified in JSL, and scripts attached to a data table can be run directly using JSL.

Example

Suppose you work for a candy company that is concerned with the increasing prices of raw materials and the impact that the increases will have on earnings. You have been asked to gather the historical trends of a particular raw material, and make this information accessible to everyone in your group. One person in your group is assigned the task of keeping the table updated with the most recent data and generating a monthly report. A script to generate a trend graph is attached to the data table.

There is also an associated handling fee that is a percentage of the cost per ton. The fee changes, so it is made into a table variable and can be easily updated. Once the value is changed in this table variable, the column values are automatically updated. The data table RawMaterialPrices.jmp is used in this example.

Important Note

In Chapter 2, you learned how to reference a JMP data table in JSL. When working with table variables and properties or scripts, the data table must be referenced explicitly. An error occurs if a data table is not explicitly referenced when setting, deleting, or modifying a table variable or property. JSL does not default to the current data table when working with table variables and properties.

The following script explicitly references the data table, deletes the existing table variable named Last Update, and adds a new table variable with the most recent date. It retrieves the script that is attached to the data table, and generates an overlay plot of monthly raw material prices versus the date. (See Figure 3.16.)

```
//Reference data table
rawmat_dt = Data Table( "RawMaterialPrices" );

//Delete the table variable in order to update the latest date
rawmat_dt << Delete Table Variable( "Last Update" );

//Find most recent date in table and add to table variable
last_update = Col Maximum( Column( rawmat_dt, "Date" ) );
rawmat_dt << Set Table Variable(
   "Last Update",
   "This table was last updated " || Format( last_update, "m/y" )
);


//Run Overlay Plot script that is attached to the data table
rawmat_trend = rawmat_dt << Get Property( "Overlay Plot" );
rawmat_trend;
```

Figure 3.16 Plot of Monthly Raw Material Prices versus Date

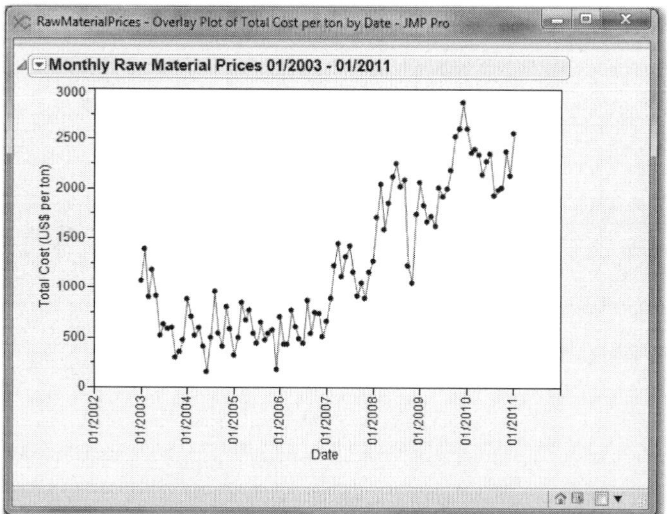

A Few Items to Note

- Variables defined in the script can be used when setting table variables.
- Date formats are available when setting table variables.
- Global variables from the main script can be used in other scripts that are run, such as the overlay plot script that is attached to the data table.
- If you are using the **Tables ▶ Summary** command, and you are repeatedly concatenating summary tables, a helpful tip is to delete the **Source** table property after each concatenation using the command datatable << Delete Table Property("Source"). When concatenating tables, the **Source** table property is added. So, if this is done repeatedly, there will be many **Source** table properties in the table, which is unnecessary.
- The **Include( )** function is an alternative to using table property scripts. The **Include( )** function parses and executes a script, and is discussed in detail in Chapter 9, "Writing Flexible Code."

4 Essentials: Variables, Formats, and Expressions

Introduction

There is no shame in using JMP to write your script for you. Advanced scripters often use this capability to generate parts of their scripts. However, you will eventually need more than the **Save Script to Script Window** option. You will realize that you need an integrated robust script to achieve what you imagine. The topics in this chapter become important as you begin your journey into writing robust, well-planned programs. You will need to use formats and formulas, conditional functions, and iterations to extend your script's functionality. You will need to control the evaluation order of your scripts. Variable management, which includes creating, scoping, and checking values and types, makes a script more robust. We think JMP dates are worthy of their own section because of the complexity and myriad of formats that we encounter. If you ask any advanced scripter about expressions, a passionate response about their importance will likely follow. After mastering these topics, your scripting imagination will spring to life!

Create Variables

A variable is a value stored in memory that can be referred to by name, and then used later. The naming convention for variables is discussed in detail in Chapter 1, "Getting Started with JSL." The naming convention is essentially that variable names must start with an alphabetic character or an underscore, and they cannot start with a numeral or special character. Variables can be local to a script or function or global to all scripts. In JMP 9, there are new ways of scoping variables, which are discussed later in this chapter and in Chapter 9, "Writing Flexible Code."

Most programming languages have the ability to assign variable names. In some languages, before a variable is used, both its name and characteristics must be explicitly declared. (For example, Visual Basic uses the Dim statement.). In JSL, this is not the case. Variables can be used without declaring them.

The most common JMP operator for assigning variables is the equal sign (=), which assigns the value in the evaluated expression on the right to the variable named on the left. The first argument (on the left) of an assignment statement must be capable of being assigned a value. (Often, the left-side argument is called an *L-value*.) For example:

```
a = 2 + 3;   //Valid assignment
5 = 2 + 3;   //Invalid assignment since 5 is not a valid variable name
```

When the second statement is run, the Log window reports an error. Changing the single = to double == converts the invalid assignment statement to a valid logical expression. JMP recognizes the statement as true, and reports a 1 in the Log window. If an assignment is needed, then you must use a single equal sign.

If by accident an assignment statement uses double equal signs, there are two possible outcomes. JMP will throw an error if the L-value has not been previously assigned. Or if the variable has been previously assigned, JMP evaluates the statement as a true or false value, does not store the value, and the script continues to run. Since the intended assignment did not occur, it is likely the script produces incorrect results. This error can be extremely difficult to troubleshoot. Other assignment operators are discussed in the "Operators" section in Chapter 1.

In the previous chapters and their examples, assignment statements have been used to refer to data tables, table columns, reports, and other JMP objects. This is done because the data table, table column, report, or other object is used later in the script, and it is easier to refer to the variable name. Assigning a variable to a JMP object allows messages to be sent directly to the object at any point in the script. Here is the general syntax for sending messages to an object:

object << message1 << message2 << ...

In the section "Create Data Tables" in Chapter 2, "Reading and Saving Data," you learned how to create a data table when values were known. In this example, the name prop_dt is assigned to the newly created data table. To get the script, a message is sent to the data table. Using the variable prop_dt to reference the data table is easier and more general than using the **Data Table( )** function.

```
prop_dt = New Table( "Proportions",
   New Column( "height (inches)", Numeric, Continuous,
       Set Values( [72, 68, 73, 72, 76, 66, 65, 71] )),
   New Column( "arm span (inches)", Numeric, Continuous,
       Set Values( [71, 68, 74, 73, 77, 65, 65, 72] )),
   New Column( "ratio height/arm span", Numeric, Continuous,
       Format( "Fixed Dec", 12, 3 ),
       Formula( :Name( "height (inches)" ) / :Name( "arm span
       (inches)" ) ))
); //end New Table

prop_dt << get script;
```

When using commands from the **Tables** menu, it is especially important to keep track of newly created data tables and assign them variable names. The behavior in JMP 9 is more forgiving when using commands from the **Tables** menu, but it is better to be safe and explicitly define the source data table to be used.

The scripts for the following examples are in 4_CreateVariables.jsl. The scripts are most informative if the lines of code are run as they are presented, rather than executing the entire script.

Example

The following example summarizes the data using **Tables ▶ Summary**, and then creates a plot of the data:

```
//Summarize Football data table and create overlay plot
//Open the Football data table and assign it a variable name
football_dt = Open( "$SAMPLE_DATA/Football.jmp" );

//Summarize the Weight of the players and assign the new table a name
fb_sum_dt = football_dt << Summary( Group( :Position ),
   Mean( :Weight ));
```

```
//When creating an overlay plot specify the data table to be used
fb_ovlay = fb_sum_dt << Overlay Plot( X( :Position ),
   Y( :Name( "Mean(Weight)" ) ), Separate Axes( 1 ) ); //end overlay
```

Send Messages

The following example shows the syntax for creating variables and sending messages to the objects they represent:

```
bigclass_dt = Open( "$SAMPLE_DATA/Big Class.jmp" );

//Create Bivariate object
splot_biv = bigclass_dt << Bivariate( Y( :weight ), X( :height ) );

//Send a message to the Bivariate object
splot_biv << FitLine;            //linear fit
splot_biv << FitPolynomial(2);   //quadratic fit
```

Evaluate Variables

The following example creates a list of continuous variables from the sample data file Cereal.jmp, and uses an **Eval** statement in the Distribution platform to create a plot of all the variables in the list. The **Eval( )** function was discussed in Chapter 3, "Modifying and Combining Data Tables." Essentially, the **Eval** function forces the execution of an argument, and then returns the results. In this case, it forces the cereal list to be evaluated, and JMP recognizes the items in the list as columns from the current data table.

```
//Using the Eval statement with variables
cereal_dt = Open( "$SAMPLE_DATA/Cereal.jmp" );
cereal_lst = cereal_dt << Get Column Names( Continuous );
Distribution( Y( Eval( cereal_lst ) ), Normal Quantile Plot( 1 ) );
```

A Few Items to Note

- When programming, it is a good idea to develop a style for naming variables. Throughout this book, variables referring to data tables end in _dt, column variables end in _col, and other objects have their own standards. This is a programming style suggestion, not a JMP requirement.
- Some scripters choose to put spaces in their variable names for clarity. This does provide additional readability, but it makes it more difficult to copy variable names. For a variable name that has no spaces, doubling-clicking on it highlights the entire name, and copying is easy. If a variable name contains spaces, which is perfectly good syntax, the entire variable must be selected (double-clicking on it only chooses the word that is clicked).

Scope Variables

Scoping is a feature that helps programmers manage variable names so that variables are distinct in different parts of the program.

For any language, a common trait is the use of patterns—patterns of speech or syntax patterns. Our first JSL class used the name dt so often to reference a data table, that some of our students thought a data table must be named dt! Another common syntax pattern is the use of an "i" to index a **For( )** loop or an array (a list or matrix). Consequently, a common programming mistake is to have a name collision when using nested **For** loops.This mistake is so common, JMP introduces the **For** loop with a solution. In the online Help, select **JSL Functions Index ► For**. The sample script wraps a **Local** function around the **For** loop, specifying which variables remain local within the function block. This is an excellent practice to help prevent name collisions and to limit the number of global variables to maintain.

Scoping includes both syntax (scoping operators) and semantics (rules for name resolution).

:x a single colon refers to a data table column
::x a double colon refers to a global variable
x an unqualified variable uses name resolution rules

The 4_Scoping.jsl script demonstrates that proper scoping prevents name collisions, even when names are indistinguishable. The following script is for illustration only, and should *never* be emulated:

```
clear globals();
delete symbols();  //start with a clean slate
::val = New Table( "i",
//table "i" is referenced by global variable ::val
  Add Rows( 10 ),
  New Column( "VAL",      //the first column is named VAL
     Numeric, Continuous, Format( "Best", 12 ),
     Set Values( [1, 2, 3, 4, 5, 6, 7, 8, 9, 10] )
   ),
  New Column( "val",      //the second column is named val
     Numeric, Continuous, Format( "Best", 12 ),
     Set Values( [-10, -9, -8, -7, -6, -5, -4, -3, -2, -1] )
   )
);  //end New Table

Local({val},  //use local variable val to loop through the table
   for(val=1, val<=nrow(::val), val++,
   show(val, :VAL[val], :val[val])
   )  //end for
);    //end Local
//all is good with proper scoping JMP knows what to do
show(::val:val[2], ::val:VAL[2]);
```

This extreme scoping example creates a global variable named val to reference a data table named i with two numeric columns named VAL and val. The script displays each column's values using a **For** loop within a **Local** function that declares the variable val as local. Run this script and check the Log window. There are no errors. JMP can interpret this jabberwocky-sounding script correctly.

The second half of the script (which is not shown here) removes scoping. As a result, JMP throws several errors in the **For** loop because of the ambiguity of the variable val.

JMP has no problem interpreting the following code because an infix colon is interpreted as table:column:

```
//JMP can handle xxx:yyy, JMP looks for table xxx and column yyy
show(val:val[2], val:VAL[2]);
```

Before JMP 9, the rules for name resolution for an unqualified variable x were the following, in the order in which they are listed:

1. If the variable is used within the context of **Function( )** or **Local( )**, then get the local value.
2. Look for a global variable with that name.
3. Look in the **Current Data Table( )** for a column with that name.

The 4_MoreScopingNamespaces.jsl script includes the following code, more examples of using **Local** declarations within functions, and introductory **Namespace** examples. Namespaces are briefly discussed in this section and in more detail in Chapter 9.

```
Clear Globals();
Delete Symbols();
x = 6;
oneFunc = Function( {a}, {Default Local}, a + x );
twoFunc = Function( {a}, {x}, a + x );
triFunc = Function( {a}, {x = -20}, a + x );

Show( oneFunc( 3 ), twoFunc( 3 ), triFunc( 3 ), x );
```

The results of the **Show( )** function can be seen in the Log window (9; . ; -17; x = 6;). oneFunc(3) returns 9 using the second name resolution rule because no local variable named x was found. twoFunc(3) returns an empty result because x is local, but has no assigned value. triFunc(3) returns -17 because a local x variable is found and assigned the value -20. External to each function, the global variable x retains its value of 6.

A namespace is new to JMP 9, and it provides another level of scoping. A **Namespace** is a collection of variable names and corresponding values. Names within a namespace must be unique, and names of namespaces must be unique. Here is the syntax to define a namespace:

```
nsRef = New Namespace( nsname, { name1 = expr1, name2 = expr2, … } );

nsOne = New Namespace ( "one", {x=11, y=12, z=13});
nsTwo = New Namespace ( "two", {x=21, y=22, z=23});

Show( x, nsOne:x, nsTwo:x); //x = 6; nsOne:x = 11; nsTwo:x = 21;
Show Namespaces();
```

Namespaces can be used as a tool to manage variable names (as shown above), or as a tool for advanced programmers to build a custom class of objects and methods. The *JMP Scripting Guide* includes a simple example that defines the namespace complex, which is a set of functions to perform complex number calculations. Predefined namespaces in JMP 9 include the following:

Global—global variables (same as in JMP 8)

Builtin—functions defined in JMP

Shared—global variables and current data table columns and variables

Here—default namespace in the executing script

Local—nearest local scope (typically with functions)

Window—namespace of the user-defined window

Box—namespace of the context box

Most novice scripters deploy namespaces to avoid name collisions. The simplest method is to enable the JMP **Here** namespace using the command Names Default To Here(1). This can be a useful feature, but we do not recommend its use until you understand what it does, and you ensure proper scoping for global variables. It is important to remember that scoping name resolution rules depends on whether the **Names Default To Here** option is enabled.

The 4_MoreScopingNamespaces.jsl script contains several introductory namespace examples and messages. Chapter 9 covers namespaces in more detail and includes scripts that demonstrate its utility.

If your scripts do not call other JSL scripts, using simple scoping syntax and local declarations might be all you need. If your scripts call other JSL scripts, then you should learn more about namespaces in the *JMP Scripting Guide* and in the section "Using Namespaces" in Chapter 9.

Check for Values and Data Types

The functions in this section are the easiest to use, and they need only a few lines to describe them. JMP calls them *inquiry functions*. They perform type checking. This section is a preface to the topic of writing robust scripts and tries to answer the question, "Why would I ever want or need to use these functions?"

JSL does not use explicit variable type declarations like other coding languages do, such as .NET. Of course, JMP has data types. The table column attribute **Data Type (Numeric | Character | Row State)** is one example where variable type is specified, and **Column Selection Filter** is another. Without formal variable type declarations, JMP does no type checking before it executes the script. Consider the following:

```
mynum = "abc";
y = sqrt(mynum);
```

JMP throws an error when this script is run, but not before. For simple scripts, type checking might not be necessary. Even for large and complex scripts, type checking should be performed selectively, and it should not be used for each command that is executed. However, if you are reading data from files, getting input from users, using functions, or have some non-modal aspects in your program where the user could delete an object, you need to add some type checking and error handling to your script. But first, study the syntax and behavior of the inquiry functions: **Is** and **Type( )**.

Boolean Is Functions

Is List( ) and **Is Number( )** are two examples of Boolean **Is** functions. These functions allow one argument. **Is Number(** arg **)** returns a 1 if arg is a number or a variable representing a number, and 0 if it is not. It's that simple. The following snippet demonstrates five **Is** functions. The results are in Figure 4.1. The scripts in this section are in 4_CheckValuesDataTypes.jsl.

```
//Simple examples for Type() and Is obj() functions
a = 4; b = "hello world";
c = {"red", "green", "blue"};
d = {0, 1, 2, 3, 4, 5, 6, 7, 8, 9};
e = [0, 1, 2, 3, 4, 5, 6, 7, 8, 9];
dt = Open( "$Sample_Data\Candy Bars.jmp" );
acol = Column( dt, 1 );
anme = acol << get name;
Show(a, Is List( a ), Is String( a ), Is Expr( a ) );
Show(b, Is List( b ), Is String( b ), Is Expr( b ) );
Show(d, Is List( d ), Is Matrix( d ) );
Show(e, Is List( e ), Is Matrix( e ) );
```

Figure 4.1 Is Function Log Window Output

```
/*:
a = 4;  Is List(a) = 0;  Is String(a) = 0;  Is Expr(a) = 0;
b = "hello world";  Is List(b) = 0;  Is String(b) = 1;  Is Expr(b) = 0;
d = {0, 1, 2, 3, 4, 5, 6, 7, 8, 9};  Is List(d) = 1;  Is Matrix(d) = 0;
e = [0, 1, 2, 3, 4, 5, 6, 7, 8, 9];  Is List(e) = 0;  Is Matrix(e) = 1;
```

To get a list of all **Is** functions, in the online Help, select **JSL Functions Index ► All ► Is**. In addition, you can reference the *JMP Scripting Guide* or, in a Script window, type Is, hold down the CTRL key, and select ENTER.

Type Function

The **Type(ref)** function returns the JMP object type for the ref argument. The possible JMP object types are Table, Column, Number, String, List, Matrix, Expression, Integer, Date, Empty, Pattern, Associative Array, Blob, Row State, Name, Namespace, DisplayBox, Picture, Null, and Unknown.

```
Show( a, Type( a ), b, Type( b ), c, Type( c ), d, Type( d ),
  e, Type( e ), dt, Type( dt ), acol, Type( acol ),
  anme, Type( anme ) );  //end Show
```

Figure 4.2 Type Function Log Window Output

```
/*:
Type(a) = "Number";  Type(b) = "String";  Type(c) = "List";  Type(d) = "List";
Type(e) = "Matrix";  Type(dt) = "Table";  Type(acol) = "Column";
Type(anme) = "String";
```

Study the following code to see the behavior of **Type( )** and **Is Empty( )** when a column is deleted, a table is closed, or a global variable is cleared. The type **Column** persists after acol, column 1, is deleted. Similarly, the type **Table** persists after dt is closed. However, **Is Empty( )** returns a 1—it recognizes that nothing is associated with acol or dt.

```
//Type remains even after acol is deleted
dt << delete column( 1 );
Show( Type( acol ), Is Empty( acol ) );

//Type remains even after the table is closed
Close( dt, NoSave );
Show( Type( dt ), Is Empty( dt ));

Clear Globals( dt, acol );
Show( Type( dt ), Type( acol ), Is Empty(dt), IsEmpty(acol));
```

Clearing global variables removes the object types associated with the global variables. The Log window shows the type **Null** for dt and acol after the global variables are cleared. It is not necessary to clear a global variable before assigning it a new value. Here, atemp changes its value and type with a new assignment statement:

```
atemp = 3*5;     show( Type( atemp ) );  //returns "Number"
atemp = "N/A";   show( Type( atemp ) );  //returns "String"
```

The next few lines look at date types and how different inquiry functions handle uninitialized variables. Is Empty(zz) returns a 1 (true). Type(zz) throws an error because zz is uninitialized. **Is Empty( )** is usually the first checking that is done. If it is not empty, then the type is checked.

```
//Note the difference in type
xx = Today();
yy = As Date( xx );
Show( Type( xx ), Type( yy ) );

//zz is uninitialized, Type(zz) throws an error
Show( Is Empty( zz ), Type( zz ) );

//You may want to use Is Empty() before checking the type
If( !Is Empty( zz ),
  Show( Type( zz ) ),
  Show( "zz is empty" ) //else
);
```

The 4_CheckValuesDataTypes.jsl script and later chapters have more examples of type checking to make your script robust. Here are a few recommendations to check data types:

- After reading data, especially if the data are expected to have a specific structure. For example: column name (like Brand, Calories, etc.), data type (numeric or character), and potentially modeling type.
- Within a function that is used by many scripts or by many users (like an add-in). Checking data types ensures that required fields are specified and that the data types meet requirements.
- For proper values in a calculation (for example, using **Is Matrix( )** and checking for the proper size before multiplying two matrices). JMP is forgiving in many calculations and returns an empty instead of throwing an error. (For example, x = [1, 2, 3, 0, -2, 5]; y = round(log(x), 3); returns [0, 0.693, 1.099, ., ., 1.609]).

Use Formulas

The JMP Formula Editor is extraordinarily powerful. It enables a user to manipulate data and strings easily. Creating a formula in JSL is simple: let JMP write the JSL code for you. Once a column formula is created, just copy the JSL code from the Formula Editor, and then paste it into the appropriate location in the script. Test the formula to ensure that it evaluates as intended.

Example

Consider the task of creating initials from names. The script for this task is 4_Formulas.jsl, and it uses the data table SignersDeclarationofIndependence.jmp. This data table contains the names of the 56 signers of the *Declaration of Independence* of the United States. Figure 4.3 shows a snapshot of 10 of the signers. The columns in the data table include the signer and his initials. The Initials column values were created using a formula, so the JSL code can be pulled directly from the Formula Editor.

Figure 4.3 Ten Signers of the Declaration of Independence

	Signer	Initials
1	Samuel Adams	SA
2	John Adams	JA
3	Josiah Bartlett	JB
4	Carter Braxton	CB
5	Charles Carroll	CC
6	Samuel Chase	SC
7	Abraham Clark	AC
8	George Clymer	GC
9	William Ellery	WE
10	William Floyd	WF

Even though determining the initials is not a complicated formula, the Formula Editor is used to create the formula, and then its JSL code is pasted into a script. Figure 4.4 (L) shows the formula for creating initials in the Formula Editor. To get the JSL code, double-click on the formula, and then copy it. Figure 4.4 (R) shows its equivalent JSL code in the Formula Editor.

Figure 4.4 Formula Editor with Formula (L) and JSL Code (R)

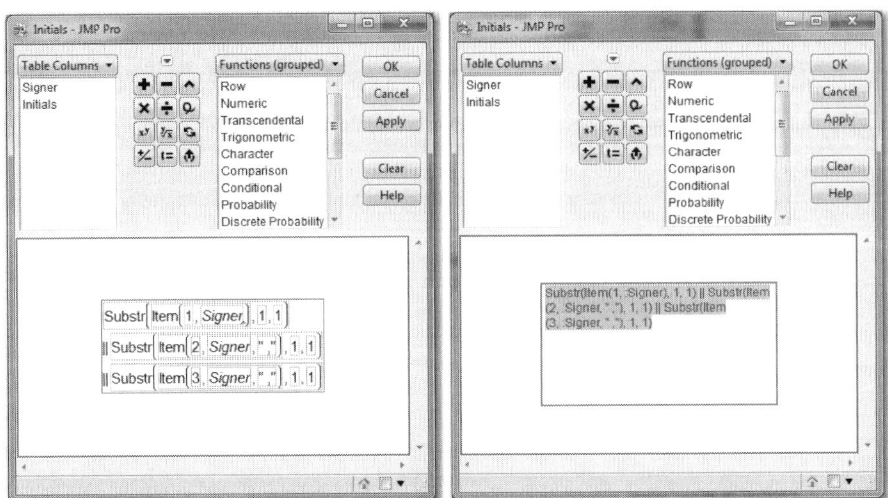

Here is the code to create the new Initials column and add the formula:

```
signers_dt = Data Table( "SignersDeclarationofIndependence" );

//---------- Add a formula when the New Column is created ---------
init_col = signers_dt << New Column( "Initials",
  Character, Nominal, Width( 5 ), Formula(
  Substr( Item( 1, :Signer ), 1, 1 )
 || Substr( Item( 2, :Signer, " ," ), 1, 1 )
 || Substr( Item( 3, :Signer, " ," ), 1, 1 ) ), //end Formula
Eval Formula );
```

The formula uses two functions: **Substr( )** and **Item( )**. **Substr( )** (the substring function) has three arguments: string, starting position, and length. It returns part of the specified string, starting at position 1 and stopping after a length of 1 (in other words, the first character). The **Item( )** function also has three arguments: integer n, string, and delimiter(s). It returns the nth substring delimited by any one of the delimiters. If a delimiter is not specified, each character is a substring. The formula concatenates the first character from the first, second, and third word of a signer's name to create the signer's initials. A formula can be added to a new column after it is created by sending it to the new column. Similarly, you can send the **Get Formula** or **Delete Formula** message to a column.

```
//---------- Add the formula to the New Column after it is created
init_col = signers_dt << New Column( "Initials",
 Character, Nominal, Width( 5 ) );
init_col << Formula(
  Substr( Item( 1, :Signer ), 1, 1 )
  || Substr( Item( 2, :Signer, " ," ), 1, 1 )
  || Substr( Item( 3, :Signer, " ," ), 1, 1 ) ); //end Formula
```

```
init_col << Eval Formula;

//----------  Get the formula and assign the formula to init_form
init_form = init_col << Get Formula;

//Delete the formula from the column
init_col << Delete Formula;
```

Control Formula Evaluation

In the example script, the message **Eval Formula** is used after a formula is created. This forces the formula to be evaluated before JMP executes another command. We highly recommend that formulas be explicitly evaluated to avoid the case where commands might be executed before the formula finishes evaluation.

An alternative to evaluating each column formula is to evaluate all formulas in a data table using the **Run Formulas** command. JMP is smart enough to execute the formulas (commands) in the correct order.

```
//Execute all formulas in a data table
signers_dt << Run Formulas;
```

There are times when suppressing a formula evaluation is needed. For example, suppose you have a very large data table, where numerous changes are expected, and the table includes multiple columns containing formulas. It might be more efficient to suppress formula evaluation using **Suppress Formula Eval( )** until the table update is completed. The following code shows how to suppress and evaluate formulas in a data table:

```
signers_dt << Suppress Formula Eval( 1 );    //formulas not evaluated
signers_dt << Suppress Formula Eval( 0 );    //formulas evaluated
```

An Alternative to Setting Formulas

Many times, retaining a formula in a column is not required. Only the values created by the formula are needed. In this case, rather than retaining the formula and having to worry about its evaluation, use the command **Set Each Value**. The syntax is identical for setting formulas and setting values. Compare the following script with the previous page. **Formula** is replaced with **Set Each Value**, and the command to evaluate the formula is not required.

```
//The syntax for Set Each Value is identical to setting a formula
init_col = signers_dt << New Column( "Initials",
  Character, Nominal, Width( 5 ) );
init_col << Set Each Value(
  Substr( Item( 1, :Signer ), 1, 1 )
  || Substr( Item( 2, :Signer, " ," ), 1, 1 )
  || Substr( Item( 3, :Signer, " ," ), 1, 1 )
);
```

Use Variables in Formulas

If your script creates a column formula that uses a variable, it is important that each variable in a formula expression is replaced with its value before the formula is assigned to the column. Otherwise, if the table was opened in a new JMP session, or the variables were cleared and the formulas were re-evaluated, the variable value is unknown, and the column does not retain its values.

Consider an example using the Candy Bars.jmp data table, where a column formula representing a linear equation is created using the column Total fat g, a slope of 8, and an intercept of 140. The values of slope and intercept are variables in the script.

When the column is created initially by the script, the formula containing the slope and intercept variables evaluates correctly. However, use caution with formulas. Once the variables are cleared, the formula loses its values if the formula is re-evaluated. To use the values in a formula rather than the variables, use the **Eval( )** and **EvalExpr( )** functions, and send the formula to the column. (See Figure 4.5.) Another option is to create table variables for slope and intercept. We prefer **Set Each Value** if the formula is not required for later use.

```
//----------  USING VARIABLES IN FORMULAS  ----------
candybars_dt = Open("$SAMPLE_DATA\Candy Bars.jmp");
cal_fat_slope = 8;
cal_fat_int = 140;

//Uses variables in the formula
candybars_dt << New Column( "Line using Total fat",
  Numeric, Continuous,
  Formula( cal_fat_int +  cal_fat_slope  * :Total fat g ) );

//Uses values of the variables in the formula
form_Expr = Expr(Column("Line using Total fat") <<
  Formula(Expr(cal_fat_int) +  Expr(cal_fat_slope)  * :Total fat g));
  Eval(EvalExpr(form_Expr));
```

Figure 4.5 Formula with Variables and Values

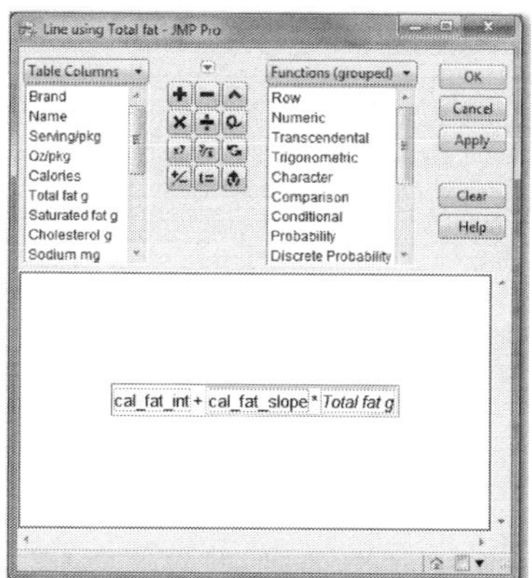

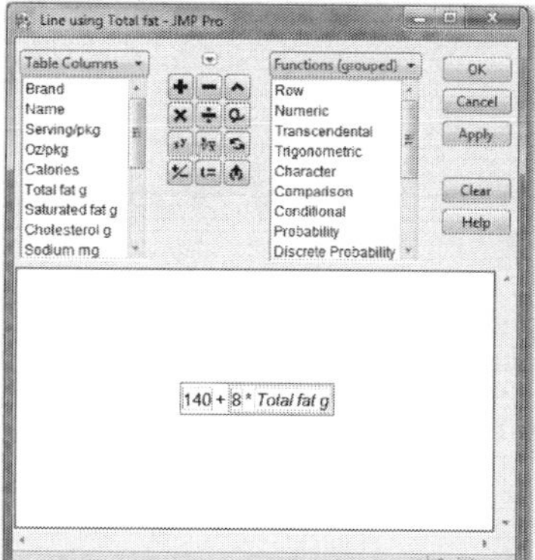

A Few Things to Note

- The **Eval( )** function works in a formula only if it is in the outermost position of the formula. Use **Substitute** or **Substitute Into** and expressions if variables are used in a formula. See the section "Expressions" in this chapter for more discussion.
- If there is a column that is referenced by another column containing a formula, and formulas are not suppressed, then the formula must be deleted before the column being referenced can be deleted from the data table. The Log window shows an error if the script deletes the column being referenced by a formula and the user's table preferences has **Suppress Formula Eval On Open** unchecked (that is, turned off).
- If calculations or built-in functions are applied to missing values, the resulting value is likely missing. Conditional formulas should be used to handle missing values if something other than a missing value is needed.

Script Timing and Execution

The default behavior of a JMP script is to execute each line or functional block of code immediately, as soon as it is read by the execution module **(Run Script)**. There are times when you need to control script timing. You might need to delay execution to ensure that system commands have completed before proceeding. Let's say, you are journaling a window and you don't want its source table to close until the journaling is complete. Also, controlling a script's timing might be required for the successful implementation of a program feature. For example, timing is important for automatically viewing multiple graphs when they are displayed one at a time, or when running simulations or demonstrations.

Typically, the system-related tasks that might require execution control are saving and closing data tables, evaluating formulas, and using platforms in which automatic recalculation is enabled.

Use Wait Function

One method to delay the execution of a script is to use the **Wait(seconds)** function. This function pauses the script temporarily for the prescribed amount of time, and then continues execution. It is especially useful when saving and closing data tables, modifying a graph, or displaying captions as a script runs. The scripts for this section are in 4_ScriptTiming.jsl.

```
For( i = 99, i > 0, i--,
  Caption(Wait( 2 ),{10, 30},
    Char( i ) || " bottles of beer on the wall, "
    || Char( i )
    || " bottles of beer; take one down  pass it around, "
    || Char( i - 1 )
    ||" bottles of beer on the wall. "
  )
);
Wait( 3 );
Caption( Remove );
```

Figure 4.6 Using the Wait Function

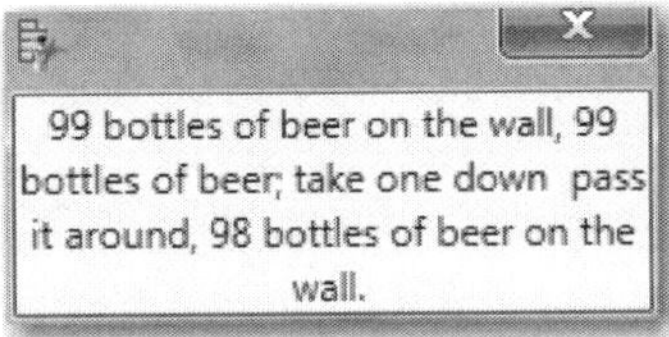

Figure 4.6 shows the first caption from the script, which uses the **Wait( )** function twice. Within the **For** loop, the **Wait( )** function delays the caption box for two seconds so that the user has time to read the caption. If that **Wait** statement is removed and the script is run, the script is executed so quickly that the only caption box that can be seen by the user is the final one. If the second **Wait** statement is also removed, then the only thing the user sees is a flash as the caption box is opened and closed very quickly.

The **Wait( )** function is used routinely to add a short delay when saving data tables that are large, or when saving them to a network path. When writing a script that will be used by many users, consider the potential size of the files and where they will be saved. Include a wait time long enough for the successful execution of the script. It is a lot easier for a user to wait a few extra seconds for a script to execute than to have the script fail because of poor timing.

A special and important use of the **Wait( )** function is **Wait(0)**, which pauses the script only as long as it takes to complete pending system tasks, and no longer.

Run Formulas

With column formulas, it is important that the formulas are completely evaluated before an analysis or summary is performed on the column or table. The **Run Formulas** message can be sent to a data table to execute all pending formulas.

```
dt << Run Formulas;
```

If the data table contains a series of formulas to evaluate, the **Run Formulas** command ensures that the formulas are executed in the appropriate order. After adding a set of formulas, JMP recommends that a **Run Formulas** command be sent to the data table

If the actual formula is not needed in the data table, which is often the case, the **Set Each Value** command can be used instead of a formula to define the values in a column. Using values instead of formulas avoids any problems with the execution and re-evaluation of columns in the data table. See the "Formulas" section in this chapter or the *JMP Scripting Guide* for more information. Several JMP platforms require column formulas, for example, the Nonlinear Modeling and the Profiler platforms. If a formula is needed for adding new data, store it as a table script that can be quickly executed.

An alternative method for controlling formula execution is to suppress the evaluation of the formula for either a specific table or globally. Then, you evaluate the formulas only when you need them. Here is the command to suppress a formula for a specific table:

```
dt << Suppress Formula Eval(1);
```

A Boolean argument of 1 suppresses the evaluation, and an argument of 0 evaluates the formulas. The following short script suppresses the formulas globally, and then evaluates them:

```
//Suppress formulas globally
Suppress Formula Eval( 1 );

new1_dt = New Table( "Random Numbers 1",
  Add Rows( 100000 ),
  New Column( "Random Normal", Formula( Random Normal() ) )
);
new2_dt = New Table( "Random Numbers 2",
  Add Rows( 100000 ),
  New Column( "Random Normal", Formula( Random Normal() ) )
);

//Evaluate formulas globally
Suppress Formula Eval( 0 );
```

Control Expression Evaluation

The section "Expressions" in this chapter is devoted entirely to writing and evaluating expressions. An expression is a container for code that is not evaluated until a separate command evaluates it. By definition, it enables a delay of code execution. Expressions help write modular code because their execution can be delayed, and then evaluated when needed.

Two commands that control the evaluation of code are **Expr( )** and **Eval( )**. The **Expr( )** command delays execution by creating a copy of its argument without evaluating it. When it is time to evaluate the argument, the **Eval( )** command can be used to execute the argument and obtain the results. The following lines of code demonstrate these two commands. Run each line of code one at a time, and then review the results in the Log window.

```
//Delay execution and evaluation of an expression
a = Expr( b + 4 );  Show( a );  //a = b + 4;
b = 1;  Show( a );  Eval( a );  //a = b + 4;  5
b = 2;  Show( a );  Eval( a );  //a = b + 4;  6
```

Conditional Functions

Ask a group of programmers for the conditional function that they use most often, and the If function is likely the winning response, with a Case, Select, or Switch statement coming in next. (In JMP, the Match and Choose functions are similar to the Case command in other languages.) Conditional functions have no standard syntax or behavior across all languages, yet they share a general purpose: the value of a Boolean variable or expression controls what is done next. After demonstrating the syntax and characteristics of conditional functions, this section's script includes extra examples using **If** and **Choose** functions to control program flow.

If Function

In the following syntax statements, condition represents a Boolean variable or expression—a variable or expression that returns a 0|1 or True|False when evaluated. In many programming languages, the If function allows only one condition and two result arguments. The If function returns the first result when the condition is True, otherwise, it returns the second result. In JMP, multiple condition-result pairs are allowed. When an odd number of arguments are specified, the last argument serves as a default or an else result for non-missing conditions.

```
If(condition, result, condition2, result2, ..., <elseResult>);
IfMZ(condition, result, condition2, result2, ..., <elseResult>);
```

Run each block of code from the 4_ConditionalFunctions.jsl script.

```
//If and IfMZ simple syntax & behavior
xx = 5;
yy = If( xx > 0, "+1", xx < 0, "-1", "0" );
Show( yy );  //yy = "+1";

xx=.;  //same as xx=empty();
yy = If( xx > 0, "+1", xx < 0, "-1", "0" );
show( yy, Type ( yy ) ); //yy = .; Type(yy)= "Number";

//Control empties with type checking
show( Type(xx), Is Empty(xx), Is Missing(xx), xx == Empty() );
//Type(xx) = "Number"; Is Empty(xx) = 0; Is Missing(xx) = 1;
//xx == Empty() = .;

yy = If( Is Missing(xx), "??", xx > 0, "+1", xx < 0, "-1", "0");
show(yy, Type(yy)); //yy = "??"; Type(yy)="String";

//IfMZ - a missing value is forced to zero
xx = .;  //xx=empty();
yy = IfMZ( xx > 0, "+1", xx < 0, "-1", "0" );
Show( yy, Type( yy ) ); //yy = "0"; Type(yy)="String";

xx = .;  //xx=empty();

//Matched condition-result pairs & last condition is always true
yy = If( xx > 0, "+1", xx < 0, "-1", 1, "0" );
Show( yy, Type( yy ) ); //yy= "0"; Type(yy)="String";
```

This portion of the script demonstrates the basic functionality of the If function. Study the cases where xx is missing—the If function's behavior might not be intuitive. The **Is Missing( )** function is the correct function to test for a numeric empty. Remember that the **Is Empty( )** function returns a 1 when a value is uninitialized. Also, remember that JMP returns empty for any calculation or function that involves an empty variable. If xx is empty, then the expressions xx > 0, xx < 0, and xx == Empty( ) all evaluate to an empty numeric value **(.)**, not a 0|1, True|False result. As a consequence, If **(empty,..)** returns an empty. The last block of code uses the constant 1 as the final condition, so 0 becomes the default value. Some programmers like this method to set default values. Add comments to your scripts when using this default-value construct because not every user is familiar with this method.

The **IfMZ** function replaces a missing value with a zero. This function provides another alternative. There are several options for handling missing values, including total control using an **IsMissing(xx)** condition and result argument. (If **IsMissing(xx)** is used, it should be placed before other condition and result arguments.)

Match Function

The **Match** function is a shortened form of the **If** function, in which each condition is an equivalency (==) expression. When the variable or expression in the first argument **(a)** matches a value, the **Match** function returns the corresponding result. The **Match** function eliminates the repetitive equivalency expressions.

```
Match(a, value1, result1, value2, result2, ...);
MatchMZ(a, value1, result1, value2, result2, ...);

xx="F";
yy = Match(xx,"F","Female", "M", "Male", "Unknown");

//identical to If() when the conditions are equivalency(==)
xx = "";
yy = Match(xx,"F","Female", "M", "Male", "Unknown");
zz = If(xx == "F", "Female", xx == "M", "Male", "Unknown");
```

Choose Function

The **Choose** function is a special case of the **Match** function. In the **Choose** function, the first argument must be an expression that evaluates to a positive integer. Its syntax needs no comparative values. In the following example, the **Day of Week** function returns an integer from 1 to 7. The **Choose** or **Match** function assigns a name to the day based on the integer. The **Choose** function eliminates the need to list the matching value options. The requirement that **a** must evaluate to an integer between 1 to *k* might seem restrictive. However, when asking users for options using buttons and other dialog box features, restricting the values between 1 to *k* is not uncommon.

```
Choose(a, result1, result2, result3, ..., rElse);

xx = Today();
yy = Choose( Day of Week( xx ), "Sun", "Mon", "Tue", "Wed",
   "Thu", "Fri", "Sat" );
zz = Match( Day of Week( xx ),1, "Sun", 2, "Mon", 3, "Tue",
   4, "Wed", 5, "Thu", 6, "Fri", "Sat");
```

Extra Examples

Three additional scripts are included in 4_ConditonalFunctions.jsl to provide more context for using conditional functions. They use commands that are presented later in this book. These additional scripts perform the following tasks:

1. Open a data table. Add new columns for control limits and a column to flag out-of-control results using an **If( )** statement. Create an overlay trend plot with the JMP 9 **Marker Selection Mode** option.

2. Open a data table of survey data in which students ranked the importance of Grades, Looks, Sports, and Popularity from 1 to 4. Use the conditional **IfMax** and **IfMin** functions to identify not the minimum value, but which variable has the minimum ranking. Do the same for maximum ranking.

3. Open a data table of process control data for diameter measurements. Prompt the user for the type of control chart to generate using buttons. Use the **Choose** function to specify which JSL code is run.

Iterate

JSL has several functions that loop or iterate. We have demonstrated some of these functions in earlier chapters without much explanation. The basic structure of an iteration loop includes the following:

- Start or initial condition
- Test or while condition
- Increment, step size, or iteration sequence
- Body or block of commands to be executed

For Each Row and Set Each Value

For Each Row and **Set Each Value** are special iteration commands, where the initial condition is Row( ) = 1, the first row of the source data table. The end condition is the last row. The increment is 1: Row( ) += 1. The test condition can be considered a while condition: Row( ) <= N Row(dt). The first three structural components are built-in. The syntax is **For Each Row (**body**)**.

Summation and Product Functions

The **Summation** and **Product** functions are special. In these functions, the main command is built into its name. The three arguments are initial value, end value (i <= 6), and increment (1). The argument or body includes what is summed or multiplied.

```
six_evens = Summation(i=1,6,2*i);  // 42 Sum of 1st 6 even integers
six_odds = Summation(i=1,6,2*i-1); // 36 Sum of 1st 6 odd integers
power3to5 = Product(i=1,5,3);      // 3*3*3*3*3 = 3^5 = 243

a = 2; b = 4; d = (b - a)/10;      //d is step size for 10 intervals
integration_approx = Summation( i=1, 10, d*( a + ( i-.5 )*d)^2);
// 18.66  Integral(x^2) = (x^3)/3 = 64/3 - 8/3 = 56/3

show(six_evens, six_odds, power3to5, integration_approx);
```

For and While Functions

For and **While** are traditional programming iteration commands. Typically, **For** is used when counting, and when the step size is consistent or the number of iterations can be computed. **While** is most commonly used when the test or while condition requires more complex logic or the increment is conditional. Every **For** command can be written as a **While** command, but the reverse might not be true.

Example

Using the previous integration_approx example, create an iteration loop to get successively better approximations. Start with 10 intervals, and each successive iteration will double the number of intervals. This first example stops after five iterations.

```
a = 2; b = 4; lastVal = 1000;
integration_approx = 0;
For( k = 1, k <= 5, k++,
  If( k == 1,
    n = 10,
    n = 2 * n
  );
  lastVal = integration_approx; //needed to check convergence
  d = (b - 2) / n;
  integration_approx = Summation( i = 1, n, d * (a + (i - .5) * d) ^
2 );
//-- show convergence
  Show (k, n, d, integration_approx, integration_approx - lastVal);
);  //end For
```

Suppose that a specific convergence criterion is wanted. The 4_Iterating.jsl script contains several options to do this. The previous code snippet already has the necessary logic: lastVal and integration_approx have starting values. The script stops when the absolute successive difference is smaller than the convergence criterion.

One variation of this script uses the convergence criterion as an expression in the test argument of the **For** loop. The following script uses a **Break( )** to stop. This example demonstrates that each of the **For** loop arguments can be a block of code, initiating multiple variables, incrementing multiple variables, or performing other commands. Recall that semicolons bind the commands together to define a block of code to be executed.

```
//Uses Break()to stop looping, an early stop when convergence is met
//Note n *= 2 is equivalent to n = 2*n
a = 2; b = 4;
For( k = 1; lastVal = 1000; integration_approx = 0; n = 10 ,
//multiple initial values
     k < 100,                          //test
     k++; n *= 2 ,                     //increment
   lastVal = integration_approx;       //body...
   d = (b - a) / n;
```

```
      integration_approx = Summation( i = 1, n,
         d * (a + (i - .5) * d) ^ 2 );
      If (Abs( integration_approx - lastVal ) < 1e-5, Break());
   );
   Show (k, n, d, integration_approx, integration_approx - lastVal);
```

If you are new to **For** loops, review the values of k and n in the Log window for each variation of this script. After the initial loop, JMP executes the increment block of code before it executes the test block of code. Programmers call that *incrementing at the bottom of loop*. For the previous code snippet, when k=7 and n=640, the convergence criterion is met, and **Break( )** stops the **For** loop before the increment block of code is run. In the full script (which is not shown here), one variation has the convergence criterion as an expression in the test block of code. So, the increment block is run first, k=8 and n=1280, and then the test block stops the **For** loop.

Without proper checks, **For** and **While** loops can iterate indefinitely, which is called an *infinite loop*. In the previous code, the added condition k < 100 ensures that the loop will terminate after 99 iterations, as long as k is incrementing by 1 each time. Most infinite loops are caused by a lack of boundary checks and by destroying boundary variables. For example, if the command k++ was deleted, and the approximation diverged, the **For** loop would iterate indefinitely or until it created a JMP exception. There are rare instances where an intentional infinite loop is useful. Typically, they are used to check on a service or object state and they include a break or a termination statement. They are beyond the scope of this book.

There might be instances where the body of code cannot or should not be executed. Consider the JMP sample data set Semiconductor Capability.jmp. Columns P1 and M1 have only one unique value—a control chart or distribution plot is a waste of graphic space. Scenarios where values are all missing or where a user-defined column list includes a non-numeric column should be found and ignored. This scenario is a good case for a **Continue** command. In this case, inform the user, send a message with a **Caption**, write a note to the Log window or add a text box to the journal files, and continue analyzing the remaining columns in the list. There might be instances where you want to get out of the loop, but you also want to continue the program. And, in more severe instances, you might want to stop the entire program. Table 4.1 describes the behavior of functions to stop iterating.

Table 4.1 Functions Used to Stop Iterating

Function	Result
Continue( )	Stops the current iteration of a loop at the **Continue** command, and begins the loop at the start of the next iteration.
Break( )	Stops running the loop at the **Break** command, and starts executing again at the first statement outside the loop.
Throw( )	**Throw(msg)** writes a message to the Log window and aborts script execution. This function is often used with **Try( )** or **If( )** to catch the problem, create a good error message, perform steps to exit safely, and then throw a message.

Continue and **Break** can be used in the body of the **For**, **While**, and **For Each Row** functions. **Throw** can be used almost anywhere.

Extra Example

The 4_Iterating.jsl script includes a block of code that has many extra features, using techniques from previous chapters (such as **Open**, **Sort**, **Summary**, lists, and functions) and future chapters. This code snippet is an example of the **Continue** function:

```
//Create a custom chart for each response with a non-constant value
 For( i=1, i <= nitems(cList), i++,
   if( Col Std Dev( column( cList[i] ) ) == 0,
     tmpStr = cList[i] || " has no variation, no charts created";
     Caption( tmpStr );                              //caption
     VListBox( Outline Box(tmpStr) ) << journal;     //journal
     show( tmpStr );                                 //Log window
     Wait( 5 ); Caption( Remove );
     Continue()
//go to the bottom of the loop, skip remaining "body" commands
    ); // end if

   cc = lot_sumry_dt << Control Chart(
     Sample Label( :lot_id ),
     KSigma( 3 ),
     Chart Col( eval(cList[i]), Levey Jennings )
   );  //end control chart
   report(cc)[OutlineBox(2)]<<journal;
   wait(0);
   cc<<close window;
 );  //end for i
```

JMP Dates

When importing or manipulating files, it seems that one of the toughest types of data to manage are data that involve dates and times. There are many variations of the format and no standards for displaying date and time. So, even within the same data file, there might be different formats.

Many software applications handle this situation by first converting the date and time to the number of seconds since a given date. *The Beginning of JMP Time* (TBJT) is midnight on January 1, 1904. The large numeric values created when manipulating dates and times are the number of seconds since TBJT.

The scripts for this section are in 4_JMPDates.jsl. From the following examples, the long value output to the Log window is the number of seconds since TBJT. For viewing and interpreting, that number is not very useful. However, for manipulating dates, it is necessary to convert a date to a numerical value. There are many operators in JMP that convert the number of seconds to something more useful. The last line of code shows how JMP recognizes dates and times before TBJT as negative numbers, and how to convert the number of seconds into another unit of time (hours, in this case).

```
Date MDY(1,1,1904);
//0   The Beginning of JMP Time

Date MDY(10,1,2010);
//3368736000   number of seconds since midnight January 1, 1904

Date MDY(10,1,2010)/InHours(1);
//935760   number of hours since midnight January 1, 1904

Date MDY(12,31,1903)/InHours(1);
//-24   number of hours since midnight January 1, 1904
```

In the last two lines of code, the **InHours( )** function is used to convert seconds into hours by dividing the total number of seconds by the number of seconds in one hour. There are other functions similar to this one that convert to minutes, days, weeks, and years.

Before proceeding with more examples of manipulating dates in JMP, there are two functions that need discussion—the **Format( )** and **Informat( )** functions. The **Format( )** function returns the number of seconds in a more readable format that is specified by the user (such as m/d/y). The **Format( )** function defines how the date is displayed. The **Informat( )** function converts a date or time character string into a JMP date value. This conversion can be considered the format of the input that JMP uses to read the date. Using the previous date (October 1, 2010), here's how the **Format( )** and **Informat( )** functions work:

```
Format( 3368736000, "m/d/y" );
//"10/01/2010"   number of seconds output in a more readable format

Informat( "10/01/2010", "m/d/y" );
//01Oct2010   JMP date format from character string

Format( Informat( "10/01/2010", "m/d/y" ), "y/m/d" );
//"2010/10/01"   use them together to read and change date format
```

Example

In manufacturing, there is an interest in determining how long a production unit stays at a particular step or operation in the process. This is referred to as *throughput time* (TPT). The data file TPT Operation12.jmp contains information about an operation (operation 12) for the month of August 2008. There are four columns. The first two columns define the unit processed and operation of interest. The other two columns are the start and completion dates and times. The goal of the analysis is to determine the units that spent longer than 5 hours at operation 12 during the month of August 2008. See Figure 4.7 for a graph of TPT, with points exceeding the goal colored red.

```
tpt_dt = Data Table( "TPT Operation12" );

//Create a new column that contains the difference between the times
tpt_col = New Column( "TPT hr:min",
Numeric, Continuous,
Formula( :Name( "Completion Date/Time" ) - :Name( "Start Date/Time"
), Eval Formula ) );

//Format the column so that it outputs hours and minutes
//Note that we want a Duration so the format is hr:m
tpt_col << Format( "hr:m", 13, -1 );

//this is needed so that the format is changed before the
//Select Where statement - if removed the points are not colored red
Wait( 0 );

//When selecting dates and times the informat statement must be used
tpt_dt << Select Where( :Name( "TPT hr:min" ) >= Informat( "12:00",
"h:m" ) );
tpt_dt << Colors( 3 );

//Overlay as shown in the text
Overlay Plot( X( :Name( "Start Date/Time" ) ), Y( :Name( "TPT hr:min"
) ) );
```

Figure 4.7 Overlay Plot of TPT

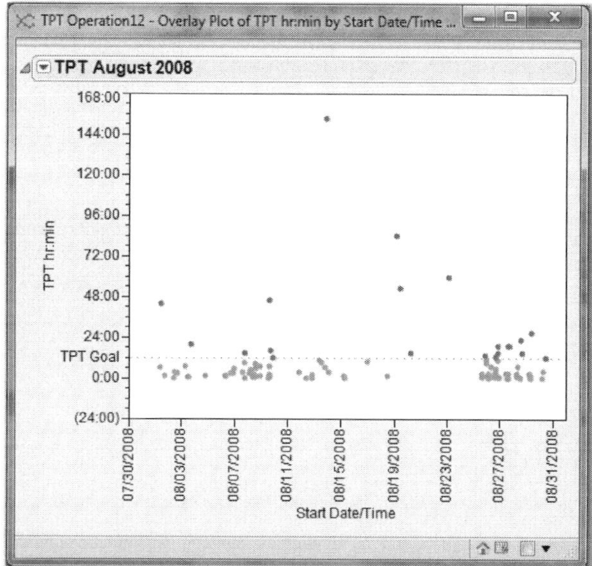

A Note About Formats

When reformatting dates and times in columns, be careful to use the right format. Some formats are similar, so ensure that the correct format is applied. An example is setting the format for tpt_col. Duration between the two times is needed, so the first line of code is used:

```
tpt_col << Format( "hr:m", 13, -1 ); //a duration format
tpt_col << Format( "h:m", 13, -1 );  //interpreted as time format
```

In the second line of code, the r has been removed from hr, and the outcome is very different. JMP sees this as a time format, rather than a duration.

Expressions

An expression is anything that can be evaluated. JMP evaluates an expression as soon as it is read by the execution module, with no delay. Often, that is exactly what you want. However, there are times in a script when immediate execution is not wanted, and control of the execution is needed. A delay might be needed because the execution of an expression is conditional on some other variable or input. The ability to control the execution of an expression is critical when writing robust and complex scripts. There are two main operators that control the execution of an expression, **Expr( )** and **Eval( )**.

The **Expr( )** function can be thought of as the delay operator. It creates a copy of its argument without evaluating it, and the argument can be reused many times and in more than one place in your script. Chapter 9 discusses the use of expressions to achieve code modularity. An expression in JMP is similar to a macro in other programming languages. The **Eval( )** function evaluates its argument, and then takes that result and evaluates it again. The functions **Expr( )** and **Eval( )** work together to delay and evaluate expressions. Table 4.2 describes these functions and other functions that are useful in evaluating expressions.

Table 4.2 Functions Used for Controlling Script Evaluation

Function	Description
Expr( )	Treats its argument as an expression, rather than evaluating it. Delays execution until the expression is evaluated. Returns its argument unevaluated.
Eval( )	Evaluates its argument, and then takes the result and evaluates it again.
Eval Expr( )	Evaluates an expression nested within an expression, and returns an expression with its argument evaluated. It does not evaluate its argument.
Name Expr( )	Returns the value of its argument unevaluated.
Substitute( ) or Substitute Into( )	Finds all matches to a pattern in an expression or list, and replaces them with another expression. **Substitute( )** evaluates its argument. **Substitute Into( )** does not evaluate its argument, but updates it in place.

Get Started

To understand these functions, it is best to start with a simple example. Run the following lines of code and note the output of each, which are marked as comments on each line. The power of expressions can be seen in the evaluation of **z** when the value of **x** is updated.

```
x = 3;
y = 4;
z = Expr(x + y);    // x + y

z;    // 7, this line and the next result in the same output
Eval(z);    // 7, same as above because it is not a nested Expr
Expr(z);    // z, its argument unevaluated
NameExpr(z);    // x + y, the value of its argument unevaluated

x = 10;    // the value of x is updated
Eval(z);    // 14, the change in x is reflected in the evaluation of z
```

Now that you understand these functions, consider other, more complex examples. Only snippets of the code are provided in this section. The full script for the examples is found in 4_Expressions.jsl.

Variables and Formulas

In the section "Formulas" in this chapter, a script is provided that shows how to use expressions so that the actual values are included in formulas, rather than variables. This is important when the information in the formula is required after variable values are cleared, such as when JMP is restarted. Consider the following example that demonstrates how to use the value, rather than the variable, in a formula. The following script uses the JMP Sample Data table Big Class.jmp. It creates a new column with the formula height + 2, using the value of the variable **bias** rather than the variable itself.

```
bigclass_dt = Open( "$SAMPLE_DATA\Big Class.jmp" );
bias = 2;
bigclass_dt << New Column( "height + bias", Numeric, Continuous );
formbias_Expr = Expr( Column( "height + bias" ) <<
Formula( :height + Expr( bias ) ) );
Eval( Eval Expr( formbias_Expr ) );
```

The next two examples use interactive displays (dialog boxes and interactive graphs), which have not yet been introduced. These are covered in Chapter 7, "Communicating with Users," and Chapter 8, "Custom Displays." However, we thought it would be useful to include these examples in this chapter because expressions are necessary when creating these types of scripts.

Dialog Boxes

There are many places where the **Expr( )** and **Eval( )** functions are needed to evaluate expressions in the right order. Arguably, one of the most important uses of these functions is when the user provides input, and then the input variables are used later in the script. The following script requests, in a dialog box, that the user inputs the numerator and denominator of the ratio. (See Figure 4.8.) It then creates a new column in the JMP data table Big Class.jmp of the ratio of the two numeric variables. The part of the script relevant to this section is the expression used to create the new column with the formula for the ratio.

Figure 4.8 Dialog Box for Computing the Ratio

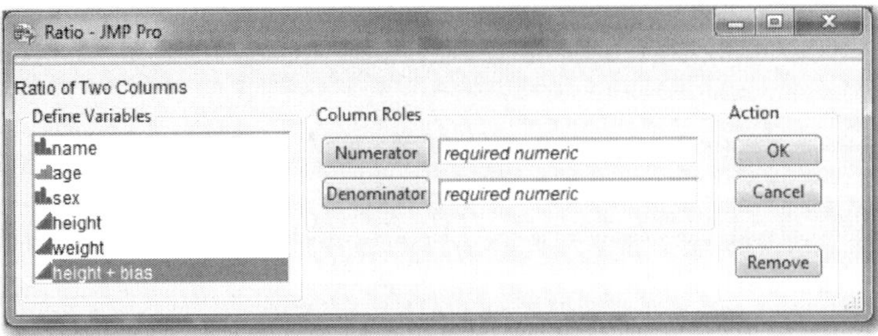

The variables numtmp_col[1] and dentmp_col[1] are the user-defined numerator and denominator values from the dialog box.

```
RatioExpr = Expr(
Ratio_col = bigclass_dt << New Column( "Ratio of " || numtmp_col[1]
|| " to " || dentmp_col[1], Numeric, Continuous );
Ratio_col << Formula( __xxx / __yyy ));

Substitute Into( RatioExpr,
  Expr( __xxx ), NameExpr(As Column(numtmp_col[1])),
  Expr( __yyy ), NameExpr(As Column(dentmp_col[1])));
RatioExpr;
```

Buttons in Interactive Graphs

Another place where expressions are necessary is in interactive graphs. If you want to include buttons that provide the user with additional options, the execution of the option needs to be delayed until the user selects the buttons. The following code provides the expression to execute a button. When the button is selected, a table with three columns is created. (See Figure 4.9.) The table consists of random Weibull data based on the parameters from the interactive graph. The following snippet of code is executed when the button is selected:

```
//Create expression for generating table of random numbers
weib_rand = Expr(
New Table( "Weibull Data", Add Rows( N ),
New Column( "Weibull(" || Char( β1 ) || ", " || Char( α1 ) || ")",
numeric, continuous, Set Each Value( Random Weibull( ___b1, ___a1 ) )
),
New Column( "Weibull(" || Char( β2 ) || ", " || Char( α2 ) || ")",
numeric, continuous, Set Each Value( Random Weibull( ___b2, ___a2 ) )
),
New Column( "Weibull(" || Char( β3 ) || ", " || Char( α3 ) || ")",
numeric, continuous, Set Each Value( Random Weibull( ___b3, ___a3 ) )
)
) );
```

```
//Substitute Weibull parameters with values from interactive graph
Substitute Into( weib_rand, Expr( ___b1 ), β1, Expr( ___a1 ), α1,
Expr( ___b2 ), β2, Expr( ___a2 ), α2, Expr( ___b3 ), β3, Expr( ___a3
), α3 );
```

The following code forces the execution of the **weib_rand** expression:

```
//Generate a data table with Weibull random data
Button Box( "Weibull Random Data", Eval( weib_rand ) );
```

Figure 4.9 Select Button to Execute Script

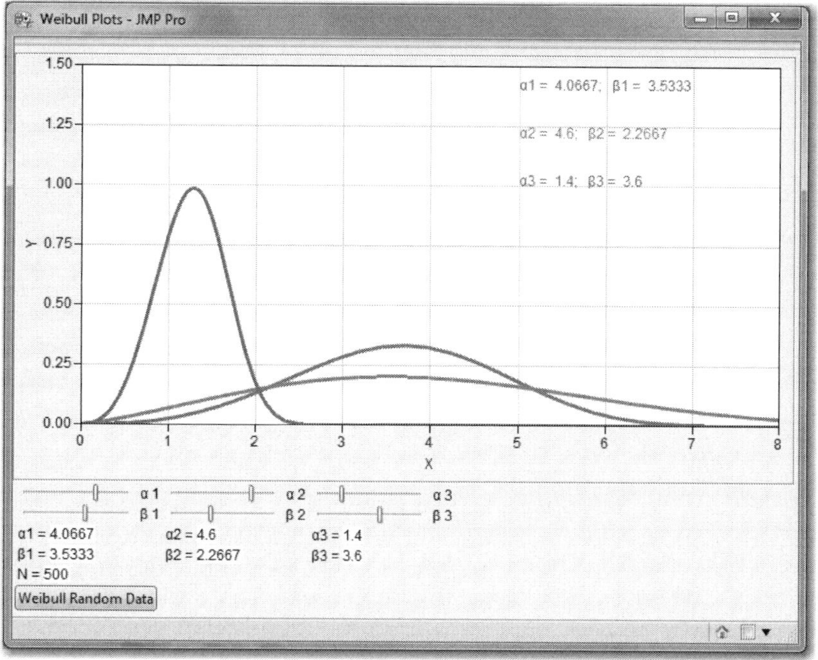

A Note About Expressions

Often, new users of JSL find expressions to be one of the more difficult topics to comprehend. The *JMP Scripting Guide* and an article by Joseph Morgan in *JMPer Cable* (Winter 2010) provide excellent discussions and examples on using expressions.

5 Lists, Matrices, and Associative Arrays

Introduction

The data structures discussed in this chapter are critical components for writing advanced scripts. Lists, matrices, and associative arrays are data structures that allow efficient storage and manipulation of various data types. Lists are pervasive in JSL. Understanding how to create and manipulate them enables the scripter to do things such as obtain input from dialog boxes and create custom output. Of the three data structures discussed in this chapter, the list is the structure most likely to be used extensively. The use of matrices makes scripts more efficient and easier to write and understand. Associative arrays are keyed or indexed lists of key-value pairs, where the key and the value can be any valid JSL object. Associative arrays are most frequently used when a script requires an efficient method to look up information.

Lists and Their Applications

After grasping the basic syntax of any JSL function or object, you should delve into its applications. Your research will probably reward you with new syntax, new JSL capabilities, and new programming concepts. Learning these concepts makes complex programs easier to write. In Chapter 1, "Getting Started with JSL," the list data structure was introduced. We included this data structure early in the book because lists are pervasive in JMP. Understanding and using them are essential to writing user-friendly, robust, and complex scripts. In JSL, information is stored in lists in a number of key places. Examples of information that is stored in lists include:

- input to a dialog box component
- a nested list of display boxes, which is the report structure of an analysis window
- graph properties such as axis settings

As important as lists are to complex programming, they are really quite straightforward. A list is a container that can store various types of JMP objects. This is in contrast to the matrix data structure, which can store only numbers. Matrices and their application are discussed in subsequent sections.

Examples and Evaluation

The items a list can store include numbers, character strings, assignment statements, functions, expressions, matrices, object references, and even other lists. Different types of objects, such as numbers, strings, or matrices, can be stored in the same list, which makes lists extremely powerful. To create a list, use the **List( )** function, or enclose the list items in curly braces { }. Items within a list, regardless of the syntax used to create the list, must be separated by commas. Examples of simple lists follow. The examples for this section are in 5_ListsandApplication.jsl.

```
a_lst = List(1, 2, 3);     //equivalent to a_lst = {1, 2, 3}
b_lst = {"a", "B", "Cc"};  //{"a", "B", "Cc"}
c_lst = {Sqrt(1), Log(5), 8*5}; //{Sqrt( 1 ), Log( 5 ), 8 * 5}
```

Generally, JMP evaluates the expression to the right of the equal sign when an assignment statement is executed. You must use expressions to delay evaluation. Lists are an exception. When the last line of code from the previous example is run, JMP returns only the list to the Log window. JMP does not evaluate the values within that list. To evaluate the items within a list, the function **Eval List( )** is required.

```
Eval List(c_lst);  //{1, 1.6094379124341, 40}
```

The previous lists are simple examples that contain items of the same type. Lists can contain items of different object types.

```
d_lst = {1, "a", sqrt(4)};  //{1, "a", Sqrt( 4 )}
Eval List(d_lst);  //{1, "a", 2}

x = "string";
y = 12;
z = Expr(bc_dt = Open("$SAMPLE_DATA/Big Class.jmp");
    bc_dt << Distribution(Y(:age)));

e_lst = {[2,2], {1,1}, x, y, z};  //{[2, 2], {1, 1}, x, y, z}
Eval List(e_lst);  //{[2, 2], {1, 1}, "string", 12, Distribution[]}
```

The variable z is an expression. When **Eval List( )** is executed, expressions in e_lst are evaluated. Consequently, the JMP Sample Data table Big Class.jmp is opened, and the Distribution platform is run using the values in column age.

Lists can contain assignment statements, functions, and even other lists.

```
//Assignment Lists
assign_lst = {a = 2, b = 3, c = 5};  //{a = 2, b = 3, c = 5}

//Before a, b, and c are assigned values, the list must be evaluated
Show(a, b, c);  //error since assignment list has not been evaluated
Eval List(assign_lst);  //{2, 3, 5}
Show(a, b, c);  //a = 2; b = 3; c = 5;

//Function Lists
mp = Function({p}, (2^p) - 1);
func_lst = {mp(2), mp(3), mp(5)};  //{mp( 2 ), mp( 3 ), mp( 5 )}
Eval List({mp(a), mp(b), mp(c)});  //{3, 7, 31}
Eval List(func_lst);  //{3, 7, 31}
mp(assign_lst);  //{3, 7, 31}
```

The mp function is able to use assign_lst as an argument because all of the values in the assignment list are numeric, and that is what the mp function requires. The last line of code—mp(assign_lst)—does not require an **Eval List( )** because the values within assign_lst are numeric. For functions to evaluate lists, all of the values in the list must be in the required format for the function. The mp function would not have executed if one of the values was a string or some other non-numeric value.

```
//This gives an error since b is a string
assign_lst2 = {a=2, b="b"};
mp(assign_lst2);  //error in log "expecting list or number"
```

Reference Items

When working with lists, it is important to know how to obtain specific values from a list. For example, if you create a script with a **Column Dialog( )**, then user selections are output as a list. You need to know how to reference the values in a list to access the user's selections. Using the lists defined previously in this chapter, you can see how subscripts are used to obtain values from lists. Subscripts can be defined by using either brackets [ ] or the **Subscript( )** function.

```
Show( a_lst[1], b_lst[2], c_lst[3] );
//a_lst[1] = 1; b_lst[2] = "B"; c_lst[3] = 8 * 5; Equivalently,
Show( Subscript( a_lst, 1 ), Subscript( b_lst, 2 ), Subscript( c_lst, 3 ) );

Show( d_lst[3], e_lst[2][1], e_lst[3] );
//d_lst[3] = Sqrt(4); e_lst[2][1] = 1; e_lst[3] = x;

Show( Eval( e_lst[3] ), Eval( e_lst[5] ) );
//Eval( e_lst[3] ) = "string"; Eval( e_lst[5] ) = Distribution[];

Show( c_lst[e_lst[1][1]] );
//c_lst[ e_lst[1][1] ] = Log(5); Equivalently,
Subscript( c_lst, Subscript( Subscript( e_lst, 1 ), 1 ) ); //Log(5)

v = assign_lst[1]; //a = 2
w = Eval(assign_lst[1]); //2
```

A Few Items to Note

- In the previous examples, the subscript of a list was called using brackets [ ]. You can also use the **Subscript( )** function. For example, Subscript(a_lst, 1) is equal to a_lst[1].
- e_lst[2][1] references the second item in e_lst and the first item within the second item.
- JMP allows nesting lists and subscripting to any level. Each Subscript must be a number except the last, the outermost, index. For example, in the expression lst[i][k][m], i and k must be numbers and m can be a number, a matrix or a list of indices.
- **Eval( )** needs to be used on individual items pulled from a list, not **Eval List( )**.
- For assignment lists, use **Eval( )** to get the value of the assignment.

Information from Lists

Lists are repositories for information. To use that information, there are several JSL functions to help extract items from lists. Extracting information from lists is a critical part of writing advanced scripts. Table 5.1 provides a quick reference to the functions discussed in this section.

For error checking, when you are expecting a list, you might first want to determine whether the input is actually a list. You can do this using the **Is List( )** function. If the argument is a list, this function returns a 1. An empty list is still a list, and a missing value is not considered a list. The following code checks for a list and if it is not a list, the code throws an error, and the script is terminated. Run the following code and replace the variable name in the **Is List( )** function for each of the four variables. The result of each variable is provided in the comments after each variable. The scripts for this section are in 5_InformationList.jsl.

```
a_lst = {1, 2, 3}; //list, script prints "Hello"
b_lst = {}; //empty list, script prints "Hello"
c_lst = .; //missing, script opens the caption
d_mat = [1, 2, 3]; //matrix, script opens the caption

//Checking for a list
If( Is List( a_lst ) != 1,
   Caption( "Script Cancelled" );
   Wait( 2 );
   Caption( remove );
   Throw() );
Print("Hello");
```

Table 5.1 List Functions

Function	Description
Is List(arg)	Returns a Boolean response that is true (1) if the argument is a list, and false (0) otherwise. An empty list is a list. A missing value is not a list.
N Items(list)	Returns the number of items in a list.
List[ ] **Subscript(list, value)**	References items from a list. (See previous section for examples.)
Loc(list, value)	Returns a matrix (column vector) of locations in the list where *value* is found.
Eval List(list)	Evaluates the items in a list. (See previous section for examples.)

Example

Using the JMP Sample Data set Cereal.jmp, lists are used to create a new column containing the abbreviations of the manufacturers, and to create Distribution plots for all continuous columns. To create a new column containing the abbreviations using JSL, there are many ways that this can be done, some of which do not use lists. (For example, **Set Each Value( )** or **Formula( )** could be used.) Because this section is on lists, this task is demonstrated two different ways using lists. Keep in mind that the second method is more efficient, and it is the recommended method for large tables. Also, the second method is faster than using either **Set Each Value** or **Formula** for large tables.

```
Cereal_dt = Open("$SAMPLE_DATA/Cereal.jmp");
Cereal_dt << New Column( "Manufacturer Abbrev", Character, Nominal,
  Width(5));
Summarize(by_var = By(:Manufacturer));
```

Both methods for creating the abbreviations column require the unique names of the manufacturers and how many there are. This information is retrieved by creating a list of manufacturers using the **Summarize( )** command, and then using **N Items( )** to determine how many there are.

In the following code, the first method loops through the list of manufacturers by using a subscript to define the items in the list, and then it applies an abbreviation to the appropriate row using an **If( )** statement and **For Each Row( )**.

```
// Method 1: Loop through the list and use For Each Row()
For( i = 1, i <= N Items( by_var ), i++,
   For Each Row(
    If( :Manufacturer == by_var[i],
     :Manufacturer Abbrev =
        Substr( Concat( Word( 1, by_var[i] ) ), 1, 1 )
        || Substr( Concat( Word( 2, by_var[i] ) ), 1, 1 ) )
   )  // end For Each Row
);   // end For i
```

The first method is slower than the second method, and it is not preferred. For large data tables, this difference is noticeable.

The second method creates a list of all values in the manufacturer column by using the **Get Values( )** command. It locates the row numbers of each occurrence of a manufacturer, and sets the abbreviation based on the row location of each occurrence of that manufacturer. The second method is faster than the first method, and is faster than using **Set Each Value( )** or **Formula( )**.

```
// Method 2: Use Get Values and set abbreviation using row numbers
Manufac_lst = :Manufacturer << Get Values;
For( i = 1, i <= N Items( by_var ), i++,
    tmp = Loc( Manufac_lst, by_var[i] );
    :Manufacturer Abbrev[tmp] =
     Substr( Concat( Word( 1, by_var[i] ) ), 1, 1 )
    || Substr( Concat( Word( 2, by_var[i] ) ), 1, 1 ) );
```

Next, create Distribution plots for all continuous columns.

```
//Create a distribution plot of the continuous columns
cont_lst = Cereal_dt << Get Column Names( Continuous );
Distribution( Column( Eval( cont_lst ) ), Horizontal Layout( 1 ),
Vertical( 0 ) );
```

A Few Items to Note

- As demonstrated in the previous example, there are multiple ways of coding tasks in JSL. When the data tables are potentially large, it is important to consider the efficiency of the code. There is more discussion about how to make your code more efficient in Chapter 10, "Helpful Tips."
- Platforms can be executed when lists are used for the variables. It is necessary to evaluate the list. See the syntax used for creating the Distribution plots in the previous script.
- The **Summarize( )** command is a powerful command for summarizing data sets without creating a new table. If a **By( )** statement is included, each variable is assigned a list of results. In this section, it found the unique list of manufacturers.
- The **Get Values( )** command retrieves values from a data table. If the data type is numeric, the output is a matrix. If the data type is character, the output is a list. The **Set Values( )** command sends values to a column in a data table. You must have the appropriate data type set for the column when sending values.

Manipulate Lists

At this point, you know the importance of lists in JSL, and you know how to extract information from them. Other important aspects of managing lists include the ability to add or remove items from a list, and to change the characteristics of these items by performing algebraic or string manipulations. As you have seen in the previous two sections, lists are powerful tools in JSL. As your skills progress, you will find that lists are key to writing programs that interact with users, create custom output, and are efficient.

Table 5.2 shows functions that enable you to insert and remove items from a list. This table is not a complete accounting of all JSL functions for manipulating lists. There are functions to sort, reverse the order, shift the items, and substitute items. These functions work on expressions as well. The **Substitute( )** and **Substitute Into( )** functions are described in the "Expressions" section in Chapter 4, "Essentials: Variables, Formats, and Expressions." There are examples for and a discussion about adding and removing items from lists in Chapter 1.

Table 5.2 List Functions for Adding and Removing Items

Function	Description
A = Insert(list, item(s), <position>)	Returns a copy of the list with the items inserted at the specified position or at the end if no position is specified. The items inserted can be a single value, a list, a matrix, or a variable. This command can be used to join lists. It does not modify the original list.
Insert Into(var, item(s), <position>)	Modifies the original list by inserting the items into the list at the specified position or at the end if no position is specified. The items inserted can be a single value, a list, a matrix, or a variable. The original list must be defined as a variable.
A = Remove(list, <position>, <#items>)	Returns a copy of the list with the items removed. By default, the number of items removed is 1 from the end of the list. It does not modify the original list.
Remove From(list, <position>, <#items>)	Modifies the original list by removing the items from the original list. It removes the specified number of items starting at a specified position. By default, the number of items removed is 1 from the end of the list. The original list must be defined as a variable.

Example

Suppose that you are interested in creating simple univariate plots (a histogram and a normal quantile plot) from a data table that contains many columns. You do not want plots of all of the continuous variables—only a subset. The JMP Sample Data table Semiconductor Capability.jmp is used to demonstrate how to do this. The script is in 5_ListManipulation.jsl. There are many ways to approach this task. One way is to create a list of all continuous columns, and then remove column names that you do not want to include. Another way is to build the list of the columns that you do want to include. In this case, let's suppose that output is needed for columns with the following attributes: columns that start with VTN and VTP, columns that start with PLY, and columns that contain the number 5. To build the appropriate list, the **If** statement uses the **Starts With( )** function, which determines whether a column name begins with a particular string. It also uses the **Contains( )** function, which determines whether a column name has a particular string in its name.

```
Semi_dt = Open( "$SAMPLE_DATA/Semiconductor Capability.jmp" );

//Create empty list - only needed if concerned about a list existing
//with the same name
cont_lst = {};

//Create a list of the variables of interest
For( i = 1, i <= N Col( Semi_dt ), i++,
tmp_col = Column( Semi_dt, i ) << Get Name();
If(
  Starts With( tmp_col, "VTN" ) |
  Starts With( tmp_col, "VTP" ) |
  Contains( tmp_col, "5" ) |
  Starts With( tmp_col, "PLY" ),
Insert Into( cont_lst, tmp_col )));

//Create a sorted list
Sort List Into( cont_lst );

//Create distribution plots with histogram and normal quantile plot
Distribution(
   Column( Eval( cont_lst ) ),
   Stack( 1 ),
   Quantiles( 0 ),
   Moments( 0 ),
   Normal Quantile Plot( 1 ),
   Horizontal Layout( 1 ),
   Vertical( 0 ),
   Capability Analysis( 1 )
);
```

The second part of this example creates a one-way analysis of the continuous columns previously defined and two nominal variables, lot_id and SITE.

```
//Create a list of nominal variables for the x-axis
nom_lst = Semi_dt << Get Column Names(Nominal, String);
Remove From( nom_lst, Loc(nom_lst, "wafer")[1]);
Remove From( nom_lst, Loc(nom_lst, "Wafer ID in lot ID")[1]);

//Create overlay plot of continuous variables vs. nominal variables
Oneway( Y( Eval(cont_lst) ), X( Eval(nom_lst) ),
   Means and Std Dev( 1 ), Box Plots( 1 ), Grand Mean( 0 ),
   Mean Lines( 0 ), Std Dev Lines( 0 ), Mean Error Bars( 0 ));
```

Algebra and Special Assignments

You can do algebraic manipulations on lists, and JSL provides some special assignments. If lists have numeric values and are conformable, then algebra can be performed on the lists.

```
a = {2, 4, 6, {1, 2}};
b = {1, 2, 3, {10, 20}};
c = a + b;  //{3, 6, 9, {11, 22}}
d = a - b;  //{1, 2, 3, {-9, -18}}
f = a / b;  //{2, 2, 2, {0.1, 0.1}}
```

In the two following lines of code where values 5 and 1 are being added to the list, the elements of the list (x, y, and z) are modified. This is unlike the line where 5 is subtracted. In this example, a new list is created, and the original list is not modified.

```
{x, y, z} = {10, 20, 30};
{x, y, z} += 5;    //adds 5 to each variable
Show( x, y, z );   //x = 15; y = 25; z = 35;
{x, y, z}++;       //adds 1 to each element
Show( x, y, z );   //x = 16; y = 26; z = 36;
{x, y, z} - 5;     //subtracts 5 from the list
```

Matrix Structure in JSL

A matrix in JMP is a mathematical construct, not the illusive 1999 science fiction cyberadventure. A mathematical matrix is also known as a rectangular array (or a two-dimensional array) of numbers. Each item in a matrix has two values to define its location: a row position and a column position. This section discusses the relationship of matrices and lists, introduces several special matrices, and provides examples of the most used matrix operations. See the *JMP Scripting Guide* and JMP online Help, available by selecting **Help ▶ JSL Functions Index ▶ Matrix**, for a complete list of matrix functions and operators. But first, you should review a matrix structure and some definitions.

Figure 5.1 Matrix Structure

$$M = \begin{bmatrix} m_{1,1} & m_{1,2} & m_{1,3} & m_{1,4} \\ m_{2,1} & m_{2,2} & m_{2,3} & m_{2,4} \end{bmatrix} = \begin{bmatrix} 10 & -25 & 22 & 0 \\ 2 & -17 & 28 & 11 \end{bmatrix}$$

Table 5.3 Matrix Definitions

Terminology	Description
Scalar	Numeric value. Operations on scalars are called arithmetic. For example, **Add(+)**, **Subtract(-)**, **Divide(/)**, **Multiply(*)**, and **Raise to a Power (^)**.
Vector	A matrix of one row or one column, also known as a one-dimensional array. JMP converts a single list of numbers to a column vector (matrix). Matrix({1,2,3,4,5}) returns [1, 2, 3, 4, 5].
Conformal	Two lists or two matrices are called *conformal* if they have the same dimensions: the same number of rows and the same number of columns.
Matrix	A matrix is a special type of list in JSL, a list that contains only numbers and has structure. Matrix arithmetic has two types of operators, *elementwise* operators and *matrix* operators. There are many matrix-only operators available in JSL. Most JMP functions applied to a matrix are applied to each element of the matrix. For example, using the matrix M in Figure 5.1, Power(M,2) = [100 625 484 0, 4 289 784 121];.

Let's start with the basics. Square brackets are used to define a matrix, and rows are delineated by commas. Run each line of the following script, one at a time, and look in the Log window for results:

```
A=[];                 // This script makes an empty matrix.
B=[2  3  5  7];       // This script makes a row vector.
C=[2,3,5,7];          // This script makes a column vector.
D=[2 3 5 7,           // This script makes a 4-by-4 matrix...
   11 13 17 19,       // ...spaces or tabs define columns...
   23 29 31 37,       // ...commas or semicolons define rows.
   41 43 47 53];
E=[+ + -,- + +,- - +]; // This makes a 3-by-3 matrix of +1s and -1s
```

JSL provides several functions to create matrices. Lists of numbers or functions can be turned into a matrix using the **Matrix( )** function.

```
//run one line at a time, see the Log
M = [10 -25 22 0, 2 -17 28 11];
M = Matrix({ {10, -25, 22, 0}, {2, -17, 28, 11} });

//reverse: Use As List() to convert M to a list
//returns a list of 2 lists,where each list is 1 row
L = As List( M );   //{ {10, -25, 22, 0}, {2, -17, 28, 11} }
```

The following matrix (**F**) is 3x4 matrix that is defined by expressions and numbers. **Matrix( )** evaluates the expressions.

```
F = Matrix({ {Sqrt(4),3,5, 1},                //row 1
             {7,11,Ceiling(pi()*4), -1},      //row 2
             {17,19,23, 0} });                //row 3
```

An identity matrix is a square matrix with ones on the main diagonal and zeros elsewhere. Identity matrices of any dimension can be created with a JSL function. In the JSL function, the argument specifies the dimension.

```
G = Identity( 5 );
```

The **J( )** function creates a matrix to your specification. The first and second arguments specify the number of rows and columns, and the third argument is the value of the elements of the matrix.

```
H = J( 2, 3, 5 );
```

The **Index( )** function generates a row vector of numbers from the first argument to the second, with an optional increment in the third argument.

```
K=Index(20,30,3);
```

The **Shape( )** function reshapes the existing matrix in the first argument into rows of the second argument and columns of the third argument. For this example if D has more than 12 elements, extra elements are discarded. If D has less than 12 elements, values will be repeated to file the specified dimensions.

```
L = Shape(D,4,3);
```

To determine whether a value is a matrix, the **IsMatrix( )** function returns a Boolean 1 if true. **NRow( )** and **NCol( )** return the number of rows and columns in a matrix or data table.

```
M = if( IsMatrix(D) & NRow(D) == NCol(D),
   Det(D),
   Caption("Determinant not calculated because the matrix is not
   square"));
```

Matrices can be added, subtracted, multiplied, and divided.

```
N = [2 3 5, 7 11 13, 17 19 23];
P = [29 31 37, 41 43 47, 53 59 61];

Q = N + P;
R = P - N;
S = N * P;
T = P / N;
```

The * and / infix operators perform matrix multiplication and matrix division when they are given two matrix arguments. They perform elementwise operations if they are given one matrix and a scalar.

To force elementwise operations on matrices, use the **EMult( )** or **EDiv( )** functions.

```
U = EMult( N, P );     // Alternatively, use U = N :* P
V = EDiv( P, N );      // Alternatively, use V = P :/ N
```

Most of the mathematical operators that work on numbers also work on matrices, and they return matrices as results.

```
W = Modulo( Q, 2 );
X = Sqrt( Q );
Y = Arctan( Q );
Z = Log( Q );
```

The **Lag( )** and **Dif( )** functions don't work on matrices or vectors, but the expected results can be easily achieved with subscripting, which is detailed in the next section.

Manipulate Matrices and Use Operations

JSL has numerous options to manipulate matrices. This section demonstrates some of the manipulations that we have found useful.

As you grow in your JSL scripting skills, you will discover many opportunities to use matrices. You might find that you need to create a matrix from a data table, or that you need to reference certain cells, rows, or columns in a matrix. Maybe you need to combine two matrices, or you need to perform some mathematics on the matrix.

So, let's concatenate some matrices! First, we'll create two 4x4 matrices of prime numbers named A and B.

```
A = [101  79 107 103, 17 89 113 13,  3 109 19 2, 59 23 61 127];
B = [ 11 131  97  67, 29 31  83 71, 47  37  7 5, 41 53 43  73];
```

Extra white space can be used between numbers to make the matrices easier to read. Now, we concatenate these matrices horizontally and vertically.

```
C = Concat(A,B);  // Makes a 4 row by 8 column matrix.
C = A||B;         // Does the same as Concat().

D = VConcat(A,B); // Makes an 8 row by 4 column matrix.
D = A|/B;         // Does the same as VConcat().
```

We can use the subscript operation to pick out elements or other matrices from existing matrices. The brackets contain the arguments for row and column, and the brackets follow the matrix to be subscripted. A single subscript treats the matrix as if all of the columns are in a single row.

```
F = C[3,4];  //Returns the value from row 3, column 4 of matrix C.
F = Subscript(C,3,4);   // Does the same thing as above.

G = D[5,3];  //Returns the value from row 5, column 3 of matrix D.
ij= [19];
G = D[ij];  // Does the same thing as above with a single subscript.

H = B[0,2];  //Returns column 2 from matrix B.
K = A[4,0];  //Returns row 4 from matrix A.
```

A subscript argument of 0 selects all rows or columns.

The *JMP Scripting Guide* demonstrates a plethora of methods to pluck values out of a matrix. We encourage you to research the many ways to extract information from matrices.

JSL provides many capabilities for solving matrix equations. Please explore the **Matrix** category of the **JSL Functions Index** for a complete list of matrix functions. Some of the more useful capabilities are demonstrated in the following example:

```
AT = A`;      //Returns the transpose of matrix A using the backquote.
AT = Transpose(A);  //Another way to transpose matrix A.

Det(B);             //Returns the determinant of matrix B.

Sweep(B);           // Returns the sweep of matrix B.

Trace(A);           // Returns the trace of matrix A.

AInv = Inverse(A);  // Returns the inverse of matrix A.

A*AInv; // Returns Identity Matrix within floating point limitations.

Round(A*AInv,15);   // Returns the Identity Matrix.
```

JSL allows you to create your own matrix operations using the **function( )** command. For instance, suppose you repeatedly need to calculate the magnitude of a vector. You could save some significant work and time by creating your own **mag( )** function.

```
mag=function({x},sqrt(x`*x));  // Create your own Magnitude function.
Mag ( H );
```

There are several JSL methods for moving information between matrices and data tables. These methods enable the wily scripter to use the efficiency and speed of matrix algebra and manipulations to perform complicated procedures, get data, and send results from and to JMP data tables.

In the following example, the Semiconductor Capability.jmp data set is opened, and the **GetAsMatrix** operation makes a matrix from all of the numeric columns in the data set. This is an example of moving a data table to a matrix.

```
SemiCapable_dt = Open("$SAMPLE_DATA/Semiconductor Capability.jmp");
M = SemiCapable_dt << GetAsMatrix;
```

As a trivial example, the arithmetic mean of the PNP1 column in the data could be calculated with the following:

```
N = V Mean( M[0,4] );
```

As an example of moving a matrix to a data table, a column vector can be moved to a data table using the **SetValues( )** operation for as many values as are in the vector.

```
Column("Prime") << SetValues(H);
```

The entire matrix can be moved with one fell swoop with the **AsTable( )** operation.

```
SemiMatrix_dt = AsTable(M);
```

Matrix Examples

Statisticians and other mathematical manipulators might already be enamored with working with matrices. But, the more casual scripters might be saying to themselves, "So what? What good is this to me?"

This section demonstrates some simple and advanced uses of matrices in examples. From these examples, we hope that the more casual scripter is motivated to use the matrix capabilities in JSL.

A simple use of matrices is to solve equations. The parameters of equations can be represented in matrix form, and then solved.

```
// Three equations; Three variables
//   x +  y +  z = 6
// 2x + 3y + 4z = 20
// 4x + 2y + 3z = 17
f = [1 1 1, 2 3 4, 4 2 3];
g = [6,20,17];
h = Inv(f)*g;  show(h);
```

The Log window returns h = [1, 2, 3] so x=1, y=2, and z=3. Isn't math thrilling?

Another excellent use of matrices and matrix algebra is to perform Ordinary Least Squares Regression. Dependent and independent variables are represented as matrices. Here is how this magic is scripted:

```
Y = [1,2,3,3,4];
X = [6,7,8,9,10];
X = J(nrow(X),1) || X;
xpxi = Inv(x`*x);
```

```
beta = xpxi*x`*y;
resid = y-x*beta;
sse = resid`*resid;
dfe = nrow(x)-ncol(x);
mse = sse/dfe;
rmse = sqrt(sse/dfe);
stdb = sqrt(vecDiag(xpxi)*mse);
alpha = .05;
qt = Students t Quantile(1-alpha/2,dfe);
betau95 = beta+qt*stdb;
betal95 = beta-qt*stdb;
tratio = beta:/stdb;
probt = (1-TDistribution(abs(tratio),dfe))*2;
show( beta, rmse, betal95, betau95, probt );
```

The Log window efficiently returns the estimated parameters, the root mean square error, 95% confidence intervals for the parameters, and the *p* values of the parameters.

Simulating or resampling is a great way to communicate probability concepts and principles to non-statisticians. JMP provides several opportunities for simulation, and JSL offers considerable power and flexibility. We've shown you how formulas and iterative loops can be used to generate simulated variables from various probability distributions in data sets for later analyses. With larger data sets, this method of simulation can take some time. The following script demonstrates an efficient method of simulation using matrices.

```
Names Default to Here(1);

// Set probability values
Pzero = 0.80; PGEtwo = 0.005; Pone = 1 - Pzero - PGEtwo;

// Set simulation size, upper control limits, zone 2 and an empty
matrix sumvalue
N = 10000; UCL = 3; z2 = 2; sumvalue = [];

// Make three matrices of zeros, all with one row and N columns.
flag = J( 1, N, 0 ); counter = J( 1, N, 0 ); supp_flag = J( 1, N, 0
);

// Set values of 0, 1, and between 2 and 7 from the probability
distribution specified above, and
// calculate the sum of those three values.
For( j = 1, j <= N, j++,
randu1 = Random Uniform();
randu2 = Random Uniform();
randu3 = Random Uniform();
value1 = If( randu1 <= Pzero, 0, If( randu1 > Pzero & randu1 <= Pzero
+ Pone, 1, Random Integer(2,7) ));
value2 = If( randu2 <= Pzero, 0, If( randu2 > Pzero & randu2 <= Pzero
+ Pone, 1, Random Integer(2,7) ));
value3 = If( randu3 <= Pzero, 0, If( randu3 > Pzero & randu3 <= Pzero
+ Pone, 1, Random Integer(2,7) ));
sumvalue = sumvalue || (value1 + value2 + value3 ) ;
);
```

```
// Fill out counter matrix based on values.
For( k = 1, k <= N, k++,
If(
sumvalue[k] > UCL, flag[k] = 99,
sumvalue[k] > z2, counter[k] = 1
);  //end if
);  // end for k

// Create a row vector of flags
comb_flag = flag + counter;

// Set comb-flag values based on flagging results.
For( m = 3, m <= N, m++,
If(
(supp_flag[m - 2] == 1) | (supp_flag[m - 1] == 1), supp_flag[m] == 0,
//reset flag after violation
((comb_flag[m - 2] + comb_flag[m - 1] + comb_flag[m] >= 2) &
(comb_flag[m - 2] + comb_flag[m - 1] + comb_flag[m] < 98)),
supp_flag[m] = 1
); // end if
); // end for k

// Make a data table and set column values from the matrices.
mean_dt = New Table( "meanvalues.jmp" );
mean_dt << Add Rows( N );
mean_dt << New Column( "Sum", values( sumvalue ) );
mean_dt << New Column( "Combined Flag", values( comb_flag ) );
mean_dt << New Column( "Supplemental Rule Flag", values( supp_flag )
);
mean_dt << New Column( "OOC", set each value( If(:Combined Flag > 1,
1, If(:Supplemental Rule Flag > 0, 1, 0)) ) );
mean_dt << Delete Column( "Column 1" );
```

This script tests how control limits and supplemental rules alarm if there is no special cause variation in a user-defined distribution of integer values. Rare events have many zeros and it can be difficult to fit a distribution. Using historical characteristics, control limits and rules can be tested for false error rates, like this example. Simulations like this can be used to estimate the average run length, or ARL.

A Few Items to Note

- Several row vectors are generated with three integer values that each range from zero to seven, and a vector is generated using the **Concat( )** operator to generate the sums of those three values.
- Several matrices are defined to keep track of flags, like value > UCL, and to count the number out of the last 3 observations where the value is > Zone2, and to reset the rules after an alarm.
- Note the subscripting used in conditional iterative loops to count the Out Of Control events, specifically the sum > UCL and the "2 out of 3 in Zone 2" supplemental control rule.

We are confident that there are more efficient methods to generate the resulting information, but we think this script demonstrates the efficiency and utility of manipulating matrices.

Associative Arrays

An associative array is a keyed or indexed list. In other programming languages, an associative array is known as a map, dictionary, hash map, or finite domain function. In JMP, an associative array is a list of key-value pairs. The unique key most often is a string or integer, and the value is any valid JSL object.

Create an Associative Array

An associative array object is initialized with one of two methods. You can use the **Associative Array( )** function, or you can use shorthand lexical syntax. Also, an associative array can be created from an existing associative array. In the JMP online Help, select **JSL Functions Index ▶ Associative Array** to get the syntax for initializing an associative array object. To get a list of valid object messages and syntax, select **Object Scripting Index ▶ Associative Array**.

A simple example of building a deli counter price list is used to introduce the associative array object. The following snippet is from the 5_AssociativeArrayBasics.jsl script. Study this script for syntax and usage. Three associative arrays (cheese, meat, and fish) are created using varying methods. The script's comments describe the method that is used. Once all three associative array objects are created, a combined product-price associative array (cppAA) is created using multiple Insert(AAobj) statements in a single command. Run this script in blocks, and look in the Log window for results.

```
//--AA created by a list of key-value pairs
cheese_AA = Associative Array({
    {"colby", 4.79}, {"swiss", 5.25}, {"havarti", 6.79},
    {"provolone", 5.79}, {"cheddar", 5.10}, {"aged_cheddar", 6.50}
    });
show( cheese_AA,  cheese_AA["aged_cheddar"]);

//--AA created by 2 same-sized lists
//  The first list defines keys, the second list defines values
meat_AA =Associative Array( {"black_forest_ham" ,"honey_baked_ham",
  "smoked_turkey", "boneless_riblets"},
  {6.99, 5.99, 5.88, 6.67}, 0  ); //meat_AA default value of zero (0)
show(meat_AA, meat_AA["smoked_turkey"]);
```

```
//--AA created with shorthand  lexical syntax
fish_AA = [ "smoked_alaskan_salmon" => 7.29,
            "teriyaki_mahi_mahi" => 8.29,
      "hawaiian_glazed_shrimp" => 6.35 , =>0 ];  //default value zero
show(fish_AA, fish_AA["teriyaki_mahi_mahi"], fish_AA["tuna"] );
//*: fish_AA["teriyaki_mahi_mahi"] = 8.29;  fish_AA["tuna"] = 0;

//--Create an empty AA, or empty AA with a default value, then insert
cppAA = Associative Array(0);  //equivalent to cppAA = [=> 0];
prices_AA = [ => ];  //equivalent to prices_AA = Associative Array();
cppAA["colby"] = 4.79;   //replaces an old or creates a new item

//syntax: Insert( key, value ) equivalent to InsertItem( key, value )
cppAA << Insert( "cheddar", 5.10 );

//----------Combine Associative Arrays------------------------------
//syntax :Insert ( AAobj)  insert an associative array object
prices_AA << Insert(cheese_AA) << Insert(meat_AA) << Insert(fish_AA);
show( prices_AA );   //prices_AA has no default value, note the order

//--AA created from two table columns, cppAA
Open( "$JSL_Companion/Deli Items.jmp" );
cppAA = Associative Array( :Item Name, :ItemPrice ); //new AAobj

//--Set a Default Value after an AAobj is created
cppAA << Set Default Value( 0 );
```

A Few Items to Note

- A key is unique. Consequently, when a key is a string, it is case sensitive.
- AAobj[key] returns the default value if the key does not exist. JMP reports an error if the key does not exist and no default value exists either.
- JMP associative arrays maintain an item order, which is the lexicographical order of the keys.
- Default values are not returned with Insert(AAobj) and AAobj << Get Contents.

The **Show** command for fish_AA["tuna"] returned 0, which is the default value. A portion of the script that is not shown includes a **Show** command for cheese_AA["pepper_jack"] and cheese_AA["Cheddar"]. The Log window reports an error similar to "subscripted key not in map {13}" when the associative array has no default value.

The default value should be chosen after considering the associative array's usage. If the default value for fish_AA was defined as a missing numeric value (=> .), every calculation involving fish_AA["tuna"] would return a missing value. The likely usage for this deli example is to multiply the number of pounds of an item purchased (tuna) by the price per pound (fish_AA["tuna"]). Producing a missing value when a non-existent item name (key) is used, or producing a zero are both reasonable options. When you want to know that a non-existent item was entered, a missing numeric is a better option. When you want to ignore a non-existent entry, zero is a better option.

Associative arrays typically have no order, not even the order in which the key-values were defined. The lexicographical order in JMP can be useful. For example, when creating a report, sorted keys can save additional coding.

When creating an associative array object from two table columns, keep in mind that keys must be unique. If an associative array is created from $Sample_Data\Big Class.jmp, and you use the command bcAA = Associative Array(:name, :age), then bcAA["Robert"] returns the age of the last Robert that appeared in the table when bcAA was created. Returning just one value for each unique item in a column can be useful. An example is provided in 5_AssociativeArrayBasics.jsl.

Remove Items

When removing items in a list, each item's location is needed. Also, when removing items from a list, you might need to start at the end of the list. For example, For(k = nitems(myList) , k > 0, k--,). An associative array's keys eliminate these requirements. The syntax to remove item(s) from an associative array is:.

```
AAobj  <<  Remove(  key |  {keys}  |  AAobj2  |  AAobj2 << Get Keys);
```

A single item, a list of items, or all the items listed in another associative array can be removed with one **Remove** command. See deli examples below.

```
//----------Remove Items---------------------------------------------
cppAA << Remove("aged_cheddar"); //remove one key
Show( cppAA << First );      //*: cppAA << First = "black_forest_ham";

//--update example
bcpp = Char( cppAA["colby"] );
cppAA["colby"] *= 1.10;    //cppAA price is updated
acpp = Char( cppAA["colby"] );
caption ("before price increase is " ||bcpp ||
      " after a 10% increase is " || acpp );

//--Remove(AAobj) example
//Note cheese_AA and cppAA have different values for key "colby"
cppAA << Remove(cheese_AA);  //same as Remove(cheese_AA << Get Keys )
show(cppAA,  cppAA << Next("smoked_alaskan_salmon"));
//*: "smoked_turkey"
```

After aged_cheddar is removed, a **Show** command displays the **First** key of cppAA. Because keys are sorted and black_forest_ham is the first key, demonstrating cppAA no longer contains an item for aged_cheddar.

Because of keyed indexing, the need to iterate through an associative array is much less than for a list. Typically, iterating is needed to cross-reference, update, or combine two associative arrays. JMP provides a handle to the **First** key in an associative array, and the command **<< Next (key)** enables looping. See the *JMP Scripting Guide*, and in the JMP online Help, select **Object Scripting Index** for a complete set of associative array messages. **Intersect** and **Next** will be of interest to you if your application requires finding common keys for two associative arrays or requires iterating.

Most modern programming languages have a data structure for a keyed array. Its name, behavior, and related functions depend on the language's implementation. In JMP, the associative array has a rich set of features: a default value; adding keys and values with subscripting; sorted keys; multiple methods for initializing; multiple object types for keys and values; and functions to manage and join associative array objects. The next section demonstrates several applications using these features.

Associative Array Applications

This section demonstrates two common uses for associative arrays: as a dictionary (also known as a lookup data structure) and as an enumerating or counting data structure.

Dictionary

Consider the task of maintaining a custom color palette for multiple reports. An associative array, where each unique color name is a key, is perfect for this task. Each key is associated (paired) with RGB (Red, Green, and Blue) settings for the color.

JMP associative arrays allow these vector values (a vector of length 3 that contains the RGB values). However, if this color palette is to be used in many scripts, it might be easier to share a table, where one column stores the color name, and another column stores the expression RGB Color(vector values). It is beneficial to have a default value, and remember that the character keys are case sensitive. For this application, black is the default color, and the keys are in lowercase.

The following script uses the key-value pairs method for defining the color palette associative array rgb_AA. The **test_color( )** function draws a swatch of color for the specified rgbName. This function uses ::rgb_AA.

```
//--Define a function that draws a swatch of the specified color-----
// Uses familiar functions Expr() and  Substitute() and
// commands discussed in later sections New Window() and Graph Box()
// Expressions and Substitute force actual values not references that
//change with each function call
test_color = Function({rgbName}, {Default Local},
drawit=Expr(New Window( rgbName,
   Graph Box(
     framesize( 150, 150 ),
     Pen Color( _rgb_);
     Fill Color(_rgb_ );
     Rect( 10, 90, 90, 10, 1); )    // end Graph Box
```

```
)); //end New Window & Expr
Swatch = Substitute( Name Expr(drawit), Expr(_rgb_),
::rgb_AA[rgbName] );
show(swatch); //after each test_color call look at the Log
swatch;
); //end Function

//--Create an associative array using a list of key-value pairs
rgb_AA = Associative Array(
  { {"tan", RGB Color([210,180, 140])},
    {"crimson", RGB Color([220,20,60]) },
    {"dodger blue", RGB Color([30, 144, 255]) },
    {"forest green", RGB Color([34,139,34]) },
    {"salmon", RGB Color([250, 128, 114]) },
    {"sienna", RGB Color([160, 82, 45]) },
    {"wheat", RGB Color([245, 222, 179]) } }
); //end rgb_AA definition

//--draw swatches of prescribed colors
test_color( "sienna" );
test_color( "forest green" );
```

The 5_AssociativeArrays_Dictionary.jsl script contains this code and several alternate commands not included here. Additional commands for adding new colors and defining a default black color follow. When storing expressions, extra precaution must be taken so that the expression is preserved, not its evaluated result.

```
//-- Add a new color using subscripting, "key" is the subscript
rgb_AA["aquamarine"]= Expr(RGB Color ([127, 255, 212]));
test_color("aquamarine");

//Add default color, JMP's color 0 is Black
//use Name Expr so default Value will not evaluate it
rgb_AA << Set Default Value( Name Expr(RGB Color(Color to RGB(0))) );

test_color("Aquamarine"); //not a valid Key draws black
```

The next commands demonstrate **Get** options. Hover or use a **Show** command to display the results. Later in the script (but not shown here), the retrieved values are used to build a table to be shared by many programs.

```
color_palette = rgb_AA << Get Contents;
color_names = rgb_AA << Get Keys;
color_rgb = rgb_AA << Get Values;
color_default = rgb_AA << Get Default Value;

//----------- How to remove a default value
//-- rgb_AA=Associative Array( rgb_AA << Get Contents);
//-- returns a list of key-value pairs, not the default value
//-- creates a new associative array with no default value
rgb2_AA = Associative Array (rgb_AA << Get Contents);
```

Extra Examples

The color palette script 5_AssociativeArray_Dictionary.jsl includes several extensions. It has code to create a table of color names and color expressions, including commands to unpack the expression for specific red, green, and blue values, and then write them to the table. It uses the associative array messages **First** and **Next(key)** to iterate. The script uses list, expression, table, and save commands to test and expand what you have learned in previous sections.

Enumerating Data Structure

Finding and counting or collecting information about unique values is another common use for associative arrays. JMP offers several methods for finding single unique values in a column.

```
//JMP provides several methods for finding a unique value
class_dt = open ("$Sample_Data\Big Class.jmp");
class_dt << New Column("ratio", numeric, continuous);
:ratio << set each value( Round( :weight/:height , 1) );

uniqA = Associative Array( :age )<< Get Keys;
Summarize( uniqB = By(:age) );
{uniqC,cumProb}  = CDF( :age << Get Values );

show( "age results", uniqA , uniqB, uniqC);
//Associative Array() numeric keys must be integers
//Summarize By returns a list of character values
//CDF only applies to numeric columns not character columns

//-- repeat for ratio
uniqA = Associative Array(:ratio) << Get Keys;
Summarize( uniqB = By(:ratio) );
{uniqC,cumProb}  = CDF( :ratio << Get Values );
show( "ratio results", uniqA , uniqB, uniqC);
```

Run these commands. The code is in 5_AssociativeArrays_Enumerating.jsl. Check the Log window for results.

Consider the task of counting or finding specific patterns or words from a column, where each row contains a long and complex text string (such as a recipe string, a shift pass-down comment, etc.).

The follows script mines the column Q7, which is a comment column from a simulated class survey. The associative array uniqQ7_AA accumulates (counts) the frequency of words. It uses shorthand lexical syntax to create an associative array with a default value of zero. It then uses the **Addto** (+=) function to find and count unique words. Each word encountered is counted. Forcing words to lowercase ensures that "Great" and "great" are counted as the same keyword. Text mining has many applications, from auditing published citations to gathering customer feedback.

```
/* Example of a simple text mining scenario using Associative Arrays
/ Q6 What did you like about this class?
/ Q7 What would you change about this class?
/ Evaluate Q7,  Modify and try it on Q6
*/
surv_dt = open("$JSL_Companion\ClassSurvey.jmp");

uniqQ7_AA=[=>0];
Local({i, j},
  for(i=1, i<=nrow(surv_dt), i++,
    words7=words(lowercase(column(surv_dt,"Q7")[i]), " .,;");
    for(j=1, j<=nitems(words7), j++,
      uniqQ7_AA[words7[j]]+=1);  //end for j
  );  //end for i
); // end Local
show( uniqQ7_AA << Get Contents );
//--------------------------------------------------------------------
//--get rid of low value words
lowValue=Associative Array({"a", "an", "and", "or", "is", "was", "i",
  "in", "of", "for", "this", "that", "these", "those", "the", "when",
  "to", "were", "has", "have"} );
nBeg = nitems(uniqQ7_AA);
uniqQ7_AA << Remove ( lowValue );  //no need to iterate…powerful!
nEnd = nitems(uniqQ7_AA);
show(nBeg, nEnd, nBeg - nEnd);
//*: nBeg = 43; nEnd = 31; nBeg - nEnd = 12;
uniqQ7_AA << Get Contents;
```

AddTo, **ConcatTo**, **AppendTo**, and **InsertInto** are accumulating (update) functions for numbers, strings, display boxes, and lists or expressions, respectively. These functions are useful in many situations. They greatly enhance the versatility of enumerative associative arrays.

Extra Examples

The 5_AssociativeArrays_TextMining.jsl script extends the previous example. It creates a traceability associative array named traceQ7_AA. Its key is a two-item list containing the respondent ID and table row number. The value for each key is an associative array like uniqQ7_AA, with the respondent's informative words as keys and counts as values. Once the array is created, the script iterates through the traceability array, and if any student's response includes focus words (such as add, boring, bored, don't, etc.), it brings up a modal view of the student's full comment. This example requires a little extra work to understand, but it demonstrates the flexibility, functions, and commands for JMP associative arrays.

6 Reports and Saving Results

Introduction

There are many books and articles that describe the techniques of visual analytics: the importance of scale and color, accentuating the necessary, stripping away the unnecessary, and more. JMP platforms create reports that are immediately useful for generating information from data. These platforms encourage you to interact with your data, explore, and learn. When it comes time to write the report, which describes and displays the important information that you have discovered, you will probably want to customize the default reports produced by JMP.

This chapter discusses JMP report and display box scripting. We approach this advanced topic by using familiar JMP platforms (Distribution, Bivariate, and Oneway) to translate point-and-click actions into JSL commands. First, we cover techniques to create and customize a single analysis report. Then we provide a rule-based, top-down method to customize multi-layered, multi-variable reports using single analysis syntax.

The critical JSL skill introduced in this chapter is learning to reference the display feature to be modified. This navigational skill begins with learning to read a map of a JMP report, which is called the JMP report tree structure. Our intention is to teach you how to access and customize components of any JMP report through examples and discussions.

An ancillary benefit of learning the JMP report tree structure is display box scripting insight and ideas for building your own custom display. These topics are covered in subsequent chapters.

We think the topics and examples in this chapter are challenging and require careful study.

Create an Analysis

Creating an analysis in JMP is as simple as a few clicks of the mouse. Customizing the report is often just as simple. Consider the point-and-click steps to create a simple linear regression for continuous variables. Open the Sample Data table Big Class.jmp. From the JMP menu, select **Analyze ► Fit Y by X**. From the dialog box (see Figure 6.1), cast the column height into the role of Y, and cast the column weight into the role of X. Click **OK**. Once the report window is created, use the pull-down menu to select the analysis option **Fit Line**.

Figure 6.1 Analysis Platforms for Fit Y by X

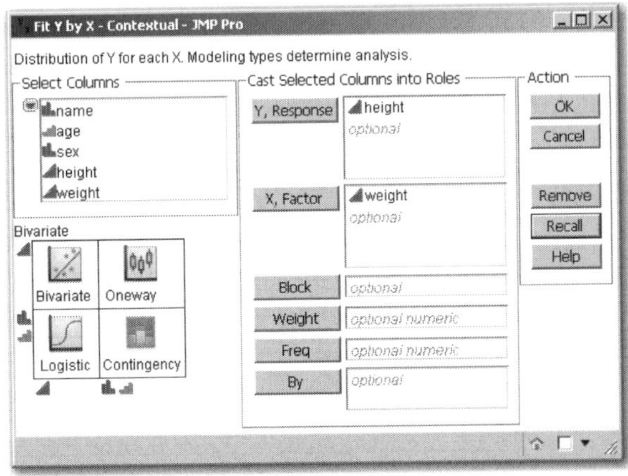

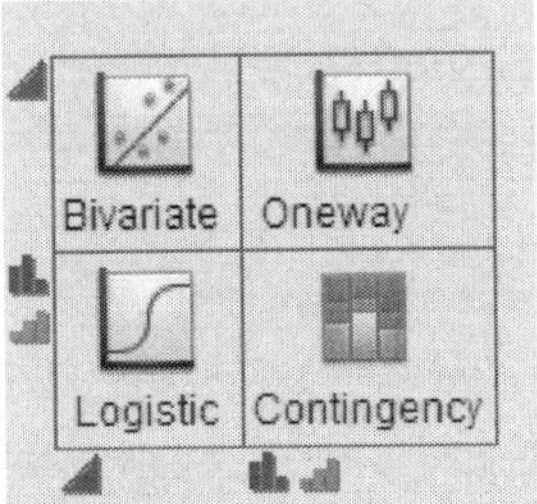

The JSL code equivalent to these point-and-click steps is the following:

```
bigclass_dt = Open("$SAMPLE_DATA/Big Class.jmp");
bigclass_dt << Fit Y by X( Y( :height ), X( :weight ), Fit Line );
//...or
bigclass_dt << Bivariate( Y( :height ), X( :weight ), Fit Line );
```

General Syntax

The general syntax for generating an analysis begins with specifying the name of the analysis platform. Then, enclosed in parentheses and separated by commas, add the arguments that define the columns cast into specific analysis roles, and add analysis items and optional actions.

```
platform_name ( Y(), … , <By( )> );
```

In most cases, the platform name matches the JMP menu name. For example, the platform names **Distribution**, **Fit Model**, and **Overlay Plot** match what is listed in the JMP menu. **Fit Y by X** is an exception to this because the analysis is context sensitive (i.e., the platform depends on the modeling type of the variables assigned to the X and Y roles). Figure 6.1 shows a larger view of the graphic from the dialog box for **Fit Y by X**. For a continuous Y column and a continuous X column, JMP runs a Bivariate analysis. Either the platform name, **Bivariate**, or **Fit Y by X** can be used.

If an analyst wants both a linear fit and a spline fit with a smoothing parameter of 1000, add analysis options to the platform command, separated by a comma.

```
bigclass_dt << Bivariate(Y( :height ), X( :weight ), Fit Line,
   Fit Spline( 1000 ));
```

Running this command produces the familiar window with graphs, summary statistics, and pull-down menus for additional graphs and analyses.

Reference the Analysis Layer

If a variable assignment is added to the platform command, that variable becomes a reference to the Analysis Layer of the platform. Sending a platform-specific message to that reference is the JSL equivalent to the point-and-click action of selecting an item in the analysis window. (See Figure 6.2.)

```
bigclass_biv = Bivariate( Y( :height ), X( :weight ), Fit Line );
bigclass_biv << Fit Spline(10000);
```

Figure 6.2 Fit Spline

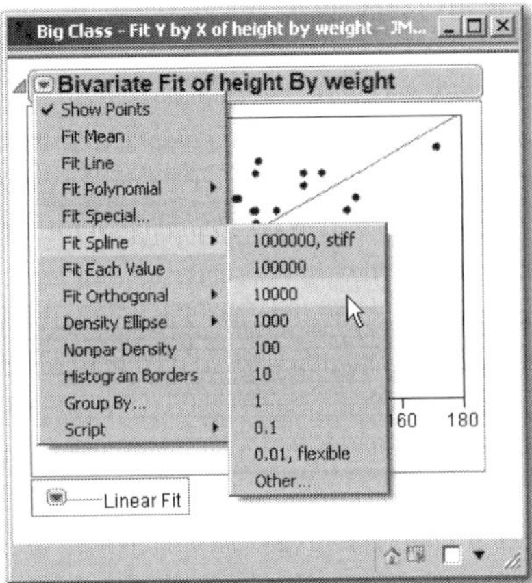

Run the following JSL statements, and review the Log window. The **Show( )** function for the reference variable bigclass_biv returns **Bivariate[ ]**, the JMP representation for a Bivariate analysis. As shown in Figure 6.3, the **Show Properties( )** function returns the list of valid messages that can be sent to this object.

```
Clear Log();
Show( bigclass_biv );  //*Log: bigclass_biv = Bivariate[];
Show Properties( bigclass_biv );
Window("Log") << Bring Window to Front;
```

Figure 6.3 Show Properties—Bivariate (repeating dots indicate items not shown)

```
bigclass_biv = Bivariate[];

Show Points [Boolean] [Default On]
Fit Mean [Action](Fits a flat line at the mean.)
Fit Line [Action](Fits a regression line to the data.)
Fit Polynomial [ActionChoice] {2,quadratic, 3,cubic, 4,quartic, 5, 6}
.................
Fit Spline [ActionChoice] {1000000, stiff, 100000, 10000, 1000, 100, 10, 1, 0.1,
0.01, flexible, Other…}(Fitting a flexible curve)
.................
Density Ellipse [ActionChoice] {0.99, 0.95, 0.90, 0.50, Other…}(The bivariate
normal contour fitted with the correlation.)
.................
Histogram Borders [Boolean]
.................
Group By [Action](Selects a grouping column so that further fits will be done on
each group.)

Curve    [Undefined] [Scripting Only]
Fit Where [Action] [Scripting Only]
```

Compare the list of properties in the Log window in Figure 6.3 with the Bivariate pull-down analysis options in Figure 6.2. The properties output lists the analysis option, and then, in brackets [ ], it lists an option such as Action, Boolean, ActionChoice, Enum, or Undefined. If the analysis option is designated ActionChoice, the available choices are in curly braces{ }. A statement inside parentheses is a menu item tip. Analysis options designated as Action do not have a standard expected value. In some instances, it might refer to other user prompts, like Fit Special.

```
bigclass_biv << Save Script to Script Window;
```

The previous line of code writes the script to the window. Find the script saved to the Script Window. Note that JMP captured the line color for each fitted curve. This is an example of a Fit Line action. The full script demonstrates other options that can be specified as an action with Fit Line (e.g., Line Width(2) and Report(0)). Thus, some customizations can be added in the same statement that generates the analysis.

```
Fit Line( {Line Color( {213, 72, 87} )} ),
Fit Spline( 1000, {Line Color( {57, 177, 67} )} )
```

JMP captures the analysis options and any changes made to graphs, such as rescaling, changing the axes, or adding reference lines. The captured script includes values determined by JMP, such as a line color or an axis scale, if the value is not specified or selected by the user.

Until recent versions of JMP, the output from **Show Properties** was an important scripting aid. Beginning with JMP 9, we recommend using the **Help ► Object Scripting Index**.

ActionChoice Messages

In the previous example, the option for adding a spline required additional information to meet the scripter's specifications. The JSL equivalent to adding information to an option is to specify the information in parentheses after the option name (for example, Spline (1000)). If the smoothing parameter of 1000 is not included, then JMP provides a default value of 100. When a required value is not specified, JMP provides a default value if it is a simple choice. The upside of JMP providing a default value is that when running the script, no coding errors occur, the user is not prompted, and the script continues. The downside is *also* that no coding errors occur. As you write and test your script, you get no messages or indications that you forgot to specify a value for the spline, and the user might see an unintended result.

If the required value is not specified for an analysis option that involves a complex choice or requires selecting columns, JMP prompts the user. Run the following script.

```
Bivariate(Y( :height ),X( :weight ),
    Fit Spline,        //no prompt default smoothing 100,
    Fit Special,       //user is prompted - choose ex. X SquareRoot
    Group by,          //user is prompted - choose column sex
    Fit Polynomial,    //no prompt polynomial 1. ie line fit
    Fit Orthogonal     //no prompt Univariate Variances, Prin Comp
);
```

The order of the **Group by** option has an impact. Only the last two fitting methods have gender-specific analyses.

Boolean Messages

The JSL equivalent to a toggle option that requires no additional information is a Boolean message. The syntax is the option name followed by parentheses with a value of 1 or 0. The value 1 is for on, and the value 0 is for off.

```
option_name( 1 | 0 )
```

You can specify Yes | No or True | False instead of 1 | 0 for Boolean messages. However, when JMP saves a script for an analysis, it uses the 1 | 0 convention, which is the same convention used throughout our example scripts.

The Oneway platform contains multiple toggle options and is used for demonstration. Consider the following example using the Sample Data table Big Class.jmp.

```
bigclass_onew = bigclass_dt << Oneway( Y( :height ), X( :age ) );
```

Figure 6.4 Oneway Analysis—No Options Specified

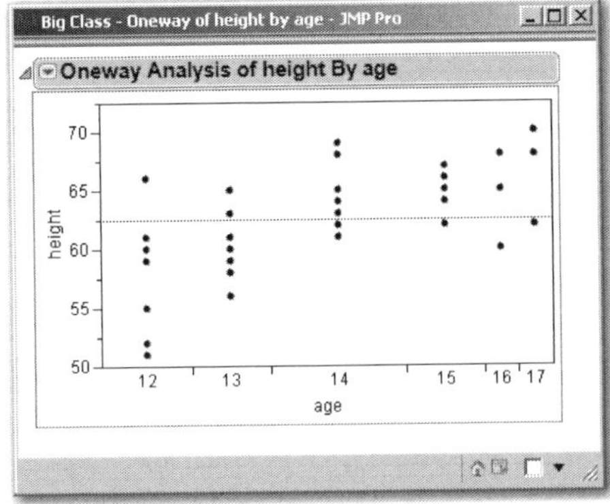

The platform command specifies no options, so the user's preferences are the only options applied to the resulting analysis. In this example, as shown in Figure 6.4, the user's options are the JMP default preferences, and a graph of only height versus age is created. This is an important feature when writing a script that will be used by other users. The scripter should always assume that different users have different platform preferences. To ensure that all users see exactly the same display, the scripter must explicitly define which options to include and exclude from an analysis.

To include additional options, send multiple messages to the analysis window using the variable name and the send **(<<)** syntax. The following first statement adds the means and standard deviations, a quantile plot, and box plots. It excludes the standard deviation lines and mean error bars from the graph. The second statement demonstrates alternate syntax that can set several options at once. X Axis Proportional is turned off, and Points Jittered is turned on.

```
bigclass_onew << Means and Std Dev(1) << Plot Actual by Quantile(1)
   << Box Plots(1) << Std Dev Lines(0) << Mean Error Bars(0);

bigclass_onew << { Means and Std Dev(1), Plot Actual by Quantile(1),
     X Axis Proportional(0), Points Jittered(1), Box Plots(1),
     Std Dev Lines(0),  Mean Error Bars(0) };
```

By and Where

When creating analyses, two useful optional arguments are **By( )** and **Where( )**. The **By( )** argument, like X( ) and Y( ), is a role that is set when the platform dialog box prompts the user. Generally, roles and other options set in a platform dialog box must be included in the JSL statement generating the platform analysis window. They cannot be sent to an existing platform.

If there is a need to restrict the data to be used in an analysis, a **Where( )** command can be specified in the JSL platform statement. The following script demonstrates two common uses for **Where** statements. The first use restricts analyses to all data of a specific subgroup, for example females. The other uses data-specific restrictions, such as younger than 16 or even formulas of variables. The **Where( )** argument is a JSL bonus. We call the **Where( )** argument a JSL bonus because it has no equivalent in the JMP menus. Either rowstates must be changed or a subset table must be created through the JMP user interface to achieve the same analysis. And, there is an even bigger bonus. The **Where( )** argument in JSL uses the original table! The platform statement can specify both **By( )** and **Where( )** arguments.

```
Oneway( Y( :height ), X( :age ), By( :sex ) );
Oneway( Y( :height ), X( :age ), Where( :sex == "F" ) );
Oneway( Y( :height ), X( :age ), Where( :age < 16  ) , By( :sex )  );
```

When letting JMP write your script and when working with a large number of By groups, it is probably more efficient to do the analysis without using a By variable. You could capture the script, and then add By(column name) to the captured script. This approach avoids the overhead of having JMP do the analysis on all of the By variables, yet still allows the user to capture the script.

Use Variable References

So far, most of the examples in this book use explicit coding: explicit column names and explicit settings. In practice, most scripts that generate an analysis use variable references. Suppose you worked for a semiconductor manufacturing company, and you are responsible for generating a C_{pk} report for 100 parameters, and then comparing this week's results with last week's. An explicit program would work until a parameter was added or deleted, but its script would be both messy to read and easy to break. A more generalized script would retrieve a list of names that could be read from a file or from the table. Or, maybe the script could prompt the user for the columns to be cast into roles for the analysis.

JMP platforms allow specifying a list of column names for specific roles. Here is a snippet from the 6_GeneratingAnalysis.jsl script. After getting a list of parameter names, a random subset of 20 names is selected. A random list emphasizes that variable names need not be pre-defined and simulates a user selection. The key point of this script is the last command. Y( ) allows a list of column names to be specified.

```
semi_dt = Open( "$Sample_Data\Semiconductor Capability.jmp" );
pnames = semi_dt << Get Column Names( Continuous );
pstr = semi_dt << Get Column Names( Continuous, String );//text names
//randomly pull 20
idx = Random Shuffle( 1 :: N Items( pstr ) );
//using pnames, column names
Show( idx[1 :: 20], pnames[idx[1 :: 20]] );
cp1 = semi_dt << Capability( Y( pnames[idx[1 :: 20]] ),
   Goal Plot( 0 ) );
```

The script also works with pstr (which is a list of the column names as text). We prefer using lists of column names as text because columns often have special characters, and using the **Name( )** syntax is avoided this way. Also, the text names have no table affiliation and are easily used in titles.

In some applications, generating an analysis through JSL is easier than the point-and-click method!

A Few Items to Note

- The 6_GeneratingAnalysis.jsl script contains the examples in this section and many more examples of generating graphs and analyses.
- The script includes several extras. There is a demonstration of storing platform preferences in a list and later applying the preferences using expressions. And, there is an advanced option to get and set a user's platform preferences.
- When capturing a script written by JMP to be reused later, the scripter should edit or remove options and settings that JMP included. These options and settings might be

appropriate for the current set of data, but not appropriate for a different set of data. Probable candidates for removal are specific **By( )** or **Where( )** groups and axis settings.

- **Where( )**, as described, can be used in any JMP platform command. The entire analysis is run on a subset of the source data table. The **Fit Where** statement (as seen in the last line in the Log window in Figure 6.3) is a very useful Bivariate scripting-only analysis option. A curve can be fit to a subset of all of the observations used in the analysis. An example is provided in this section's script and in the JMP online Help, available by selecting **Help ▶ Object Scripting Index ▶ Bivariate ▶ Fit Where**.

The Report Layer

The Report Layer is the display created by an analysis. The commands and messages from the previous section are sent to the Analysis Layer. Recall that analysis messages are the JSL equivalents to the analysis pull-down menu options.

Each platform generates a unique display, but there are common features and common characteristics in the report layout. The most familiar feature in a JMP report is the outline box. Every JMP user has probably edited an outline box title or toggled the button to show or hide the graphs and analyses belonging to that outline box.

Now that you have learned the basic syntax for generating an analysis and adding options available from the analysis pull-down menu options, the next step is to learn the JSL equivalents of the user interface tasks, like changing the outline box title or toggling a reveal button ◢.

There are two specific skills required to edit the Report Layer. One is using the valid JSL messages and syntax appropriate for the targeted display feature. The other is finding the address of the feature to be changed. We call this last skill navigation. The scripts for this section are in 6_ReportLayer.jsl.

Show Tree Structure

Navigation almost always begins with a map. Run the following script to generate a report. Then, right-click on the reveal button ◢ to the left of the outline box titled "Oneway Analysis of height By sex." Select **Edit ▶ Show Tree Structure**. (See Figure 6.5.)

```
bigclass_dt = Open("$SAMPLE_DATA/Big Class.jmp");
bigclass_ht_onew = bigclass_dt << Oneway(
  Y( :height ), X( :sex ),
  Means and Std Dev( 1 ), Mean Error Bars( 0 ),
  Std Dev Lines( 0 )
);
```

Figure 6.5 Edit Menu—Show Tree Structure: (L) Entire Report (R) Y Axis

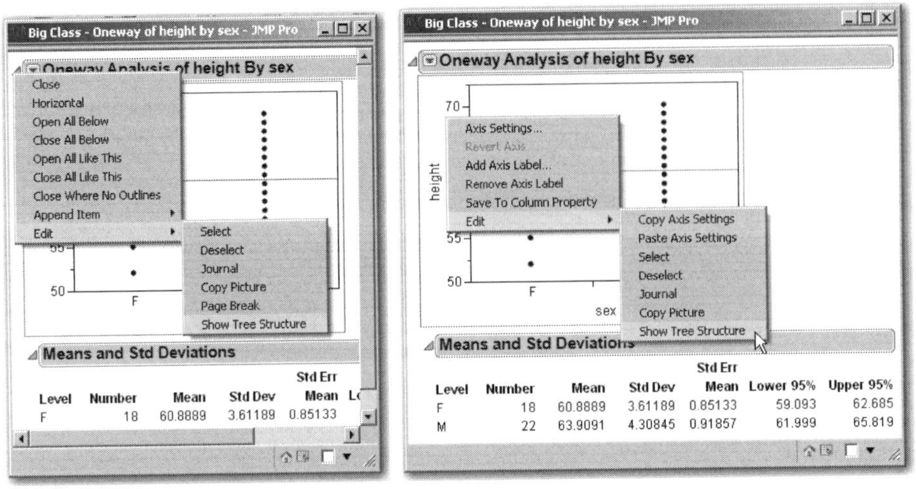

The tree structure is shown in a new window. Because the top-most reveal button was selected, the tree structure for the entire report is shown. You need to scroll to see the full tree structure. Each reveal button in the report window has the same user interface. This enables the user to show the tree structure for a smaller portion of the report window. Figure 6.5 shows that right-clicking on a report window feature like the Y axis displays a menu (**Edit ► Show Tree Structure**).

Close the tree structure for the entire report. Right-click on the reveal button to the left of the outline box titled "Means and Std Deviations" and create the tree structure.

Figure 6.6 Oneway Analysis Means and Std Deviations Tree Structure

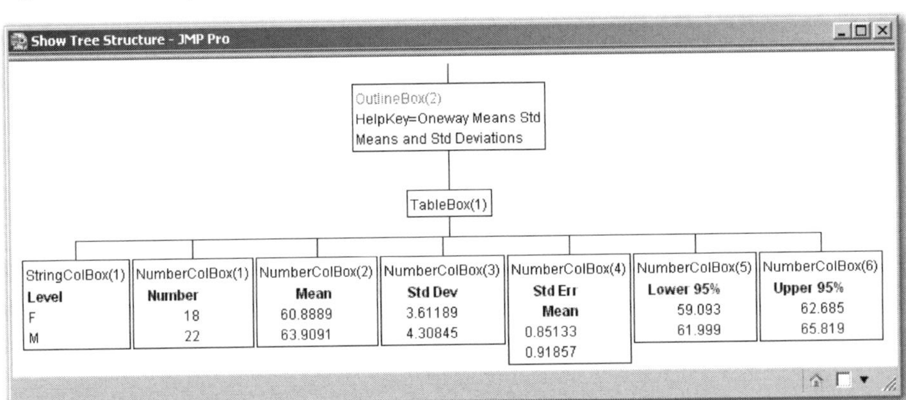

Figure 6.6 displays the objects used to arrange the output. Each object has the word "box" in its name (for example, OutlineBox, TableBox, StringColBox, and NumberColBox. The arrangement of the objects is a nested, hierarchical structure reminiscent of an organizational (org) chart.

Like a company org chart, the tree structure displays the relationships of objects in the report. And, like an org chart, the nomenclature for membership and ownership conveniently describes the display. For example TableBox(1) belongs to OutlineBox(2) and TableBox(1) owns NumberColBox(3). Later in this chapter, a parent, child, and sib relationship is described ("sib" is short for "sibling").

Reference the Report Layer

Recall that a change to the Analysis Layer requires that a valid message be sent to the analysis reference variable. For the current example, that variable is bigclass_ht_onew.

Similarly, to change the Report Layer, a valid message must be sent to the variable that is assigned to the report. The two ways to assign a variable to the report are the following:

```
rpt_bc_ht_onew = Report( bigclass_ht_onew );
rpt_bc_ht_onew = bigclass_ht_onew << Report;
```

Actually, this is the first step for most changes. To change a display feature, the valid message must be sent to the display object's address. The **Report** command returns an address for the targeted display object. For both of the previous examples, rpt_bc_ht_onew stores the address for the entire Oneway Analysis report. Using the report address and the addresses from the tree structure, and assuming the valid messages for an outline box match its JMP menu options (see Figure 6.5 (L)), this next set of statements should be apparent:

```
bigclass_ht_onew << Bring Window to Front;
//line up this script and this window
rpt_bc_ht_onew[OutlineBox(1)] << Close;
rpt_bc_ht_onew[OutlineBox(1)] << Open All Below;
rpt_bc_ht_onew[OutlineBox(1)] << Close(1);
rpt_bc_ht_onew[OutlineBox(1)] << Close(0);
```

Each display object (OutlineBox, TableBox, StringColBox) belongs to a broader classification called display box. The name OutlineBox (or NumberColBox, etc.) is an object's class name. A critical skill to learn is traversing the tree structure, which is discussed in the section "Navigate a Report" in this chapter.

JMP makes it easy to find valid messages for most display box objects. From the JMP menu, select **Help ▶ DisplayBox Scripting Index ▶ OutlineBox**. First, notice the messages that are common with the list of menu options shown in Figure 6.5. Find the message Close. Its description states that this message has a Boolean argument. Look at the sample scripts for **Get Title** and **Set Title**, and run the following JSL statement:

```
rpt_bc_ht_onew[OutlineBox(1)] << Set Title("Limit Mousing via JSL");
```

Congratulations! You've just learned the JSL equivalents to reveal or hide and to change the title of an outline box.

We offer one more useful skill before leaving this section: the ability to select and deselect parts of the report layer using the messages **Select** and **Deselect**. From the JMP menu, select **Help ▶ DisplayBox Scripting Index ▶ OutlineBox ▶ Select** and **Deselect**. Using these messages is a trick to test whether the address for a display box is correct before changing the display box. These two messages are available for every type of display box. The point-and-click equivalent is the selection tool (the fat plus sign) in a display. Here is a snippet from 6_ReportLayer.jsl with scripted delays to see the selection process:

```
rpt_bc_ht_onew[TableBox(1)] << Select;
wait(2);
rpt_bc_ht_onew[TableBox(1)] << Deselect;

rpt_bc_ht_onew[TableBox(1)][NumberColBox("Std Dev")] << Select;
wait(2);
rpt_bc_ht_onew[TableBox(1)][NumberColBox("Std Dev")] << Deselect;
```

The last pair of **Select** and **Deselect** statements is a preview of report navigation, which is described in the section "Navigate a Report." The location of the display box to be selected can be found by starting at the top of the report, and then moving down to TableBox(1). Within the TableBox(1) subordinates, you will find the NumberColBox labeled Std Dev.

If an object has a unique name or number in the report, the object can be referenced directly. The next set of commands reference, select, and deselect the same NumberColBox by its unique name and number.

```
rpt_bc_ht_onew[NumberColBox("Std Dev")] << Select;
wait(2);
rpt_bc_ht_onew[NumberColBox("Std Dev")] << Deselect;

rpt_bc_ht_onew[NumberColBox(3)] << Select;
wait(2);
rpt_bc_ht_onew[NumberColBox(3)] << Deselect;
```

Generating a report and finding valid messages for display box objects can be straightforward.

Navigating a simple report can be straightforward as well. But, there is more to learn, and navigation can get tricky when the report is more complex or when the analysis involves many Y, X, and By columns.

The next section, "Display Box Scripting," recommends a reading list to tackle the long list of display boxes and their messages.

A Few Items to Note

- The 6_ReportLayer.jsl script includes the JSL statements from this section, and provides a peek at a few navigation rules applied to this section's example rpt_bc_ht_onew.
- A user can add a custom menu to an outline box. Run an example script, available by selecting **Help ▶ DisplayBox Scripting ▶ OutlineBox ▶ Set Menu**. Select a beep or two from the hotspot pull-down menu. This capability is new in JMP 9. This feature is useful for custom dialog boxes and interactive displays.

Display Box Scripting

The previous section, "The Report Layer," provided the basic skills for customizing a report with navigation and display box scripting. This section continues the discussion of what a display box is, and targets a few common objects based on frequently asked questions.

Table 6.1 Display Box Scripting Reading List

Role	**Display Box** (Bolded items are commonly used for customizing a report)
Data	**NumberColBox, StringColBox, TableBox, TextEditBox, NumberBox, MatrixBox, TextBox**
Structure and Layout	**OutlineBox**, ListBox (VListBox and HListBox), PanelBox, LineUpBox, IfBox, TablistBox, SpacerBox, BorderBox, ScrollBox
Interactive Displays	CheckBox, RadioBox, ButtonBox, ListBoxBox, ColListBox, NumberEditBox, TextEditBox, GlobalBox, PopupBox, ComboBox, SliderBox, DataEditBox
Graphics	**AxisBox**, **FrameBox**, PictureBox, LegendBox, GraphBox, ScriptBox, JournalBox, CatAxisBox, **NomAxisBox**, WebBrowserBox, PlotColBox, CrossTabBox, CellPlotBox

Consider the display box objects listed in Table 6.1. The word "Reading" in the table title implies trying the JMP sample scripts for these messages. Bolded objects are probably required for customizing a JMP report. The following display box messages should be considered required reading for this chapter.

NumberColBox

For a NumberColBox, common customizations include messages to Set Format, Set Values, Get As Matrix, Get Heading, Set Heading, Hide, or Unhide.

OutlineBox

The tasks Set Title and Close(1 | 0) for an outline box were covered in the last section. Get Title is another widely used task. The retrieved JMP title can be used as a seed for a new title or as a check that the expected structure was selected.

Besides formatting numbers and changing titles, adding a row legend, resizing a graph, customizing a curve, and customizing axes and their labels are the most frequent requests that we encounter.

AxisBox

Novice scripters should review all available messages for AxisBox. Messages recommended for more experienced scripters include Scale, Interval, and Save to Column Property. Scale includes several probability scales, which enables the creation of custom graphs with the same characteristics as probability graph paper. Interval makes it easy to scale a time axis by year, quarter, month, second, and more.

Often, an analysis is performed on all data, and then on subsets of data based on some criteria. "How do I keep the axis the same for all subsets?" is a common question. Perform the aggregated analysis first, and then either use JMP default scaling or apply your own scaling algorithm to the subsets. Send the message Save to Column Property to the AxisBox. All subsequent graphs will use the same scaling without the JSL overhead.

Customize a Curve

After you have mastered the previous common display box objects, the next question that you will likely face is, "How do I customize a curve on a graph?" The valid messages to change a curve on a Bivariate graph are available in JMP online **Help** by selecting **Object Scripting Index**, not **DisplayBox Scripting Index**. When a message is sent to the Analysis Layer that creates an object with its own menu of actions, it is probably listed in the Object Scripting Index. Bivariate Curve, Profiler, and K Means Cluster are just a few examples of objects. See Figure 6.7 for the message syntax to edit a Bivariate Curve.

Figure 6.7 Bivariate Curve Object Scripting Index

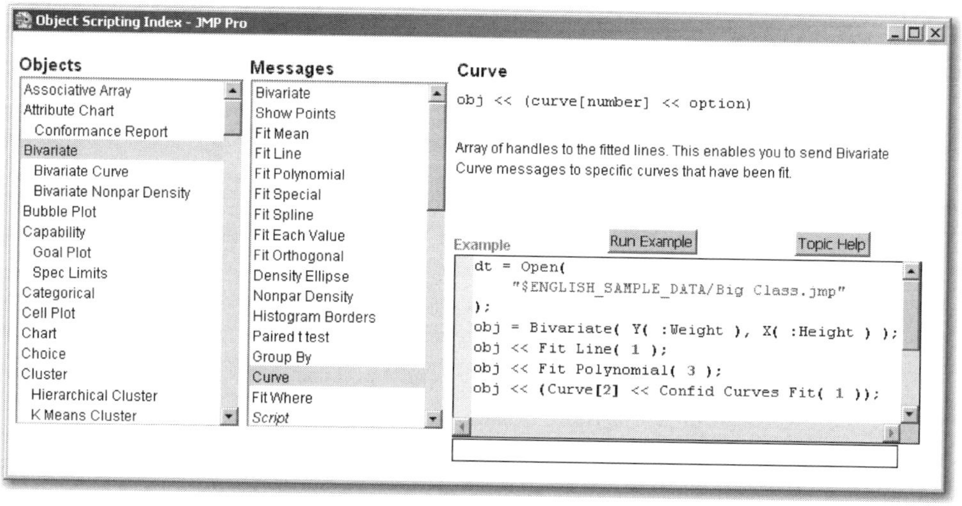

Examples

Figure 6.8 (L) displays the customized report created by the 6_ChangeReportsBivariate.jsl script. Here are some features of this report:

- Demonstrates sending messages to the platform (Analysis Layer).
- Accesses and customizes three curves, and shows how to remove a linear fit and add a third order polynomial fit.
- Customizes the OutlineBox, AxisBox, and TextEditBox (axis titles).
- Adds a LegendBox that demonstrates how to maintain the current table rowstates. Comments on how to use custom colors if no rowstates are defined.
- Using Select and Deselect, shows an example of relative and absolute display box referencing, which is discussed in subsequent sections.

Figure 6.8 Examples of Customized Reports: (L) Bivariate and (R) Oneway

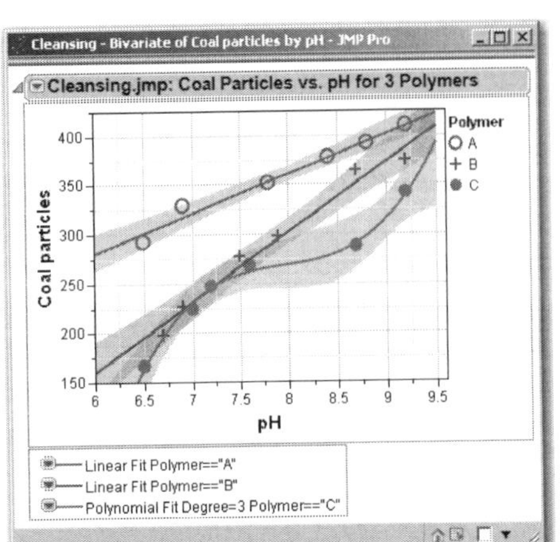

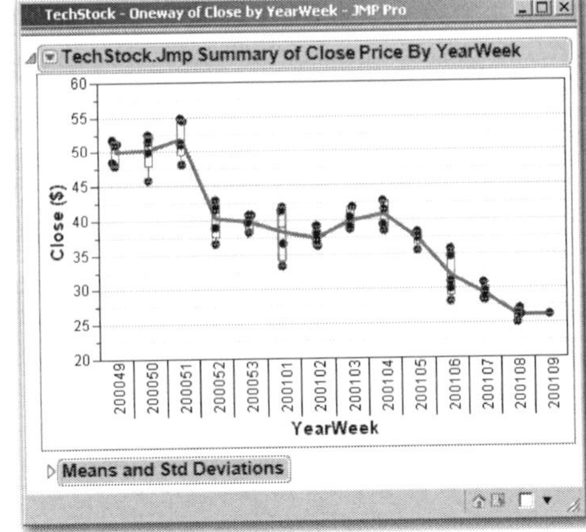

Figure 6.8 (R) is the final display after running the 6_ChangeReportsOneway.jsl script. Here are some features of this script:

- Demonstrates sending messages to the platform (Analysis Layer).
- Customizes the OutlineBox, AxisBox, and TextEditBox (axis titles).
- Highlights differences in the relationship of AxisBox to its axis label (TextEditBox) versus the NomAxisBox, which does not have the same relationship with its label.
- Customizes the Means and Std Dev TableBox by hiding several columns and formatting other data columns.
- Customizes the FrameBox by resizing, demonstrates editing one workweek's box plot using DispatchSeg, and customizes the connect line with DispatchSeg.
- Select and Deselect are used for practice navigating the display.

This entire script should be run so that you can watch the changes being made, followed by a step-by-step script review to get familiar with the syntax.

A Few Items to Note

- The 6_DisplayBoxes_AxisBox_GetStats.jsl script demonstrates adding a reference line and retrieving statistics from a regression analysis.
- Some items in the **DisplayBox Scripting Index** do not have associated scripts, and they are not documented in the *JMP Scripting Guide*. The *JMP Scripting Guide* cautions that "if you see a display box listed in the DisplayBox Index help that is not listed in the JSL Functions Index, then you can send messages to it, but you cannot create one." For example, the **TabPaneBox**.

Navigate a Report

Navigating a report is the essential skill to create a reference to a display box object. Once the location of the report component is identified, a message can be sent to modify the display box object or to retrieve information. Although the display box layout is unique for each JMP platform, messages are identical for each display object with the same class name.

Suppose the task is to customize an axis label to have a font point size of 11. The general syntax is the same, regardless of the platform:

```
path_of_label_of_Yaxis_of_Bivariate << Set Font Size( 11 );
path_of_label_of_Yaxis_of_Distribution << Set Font Size( 11 );
```

Navigation Path Syntax

The 6_Navigation_dboPathSyntax.jsl script is used to investigate the basics of JMP report navigation. Figure 6.9 contains a subsection of the tree structure generated by the following JSL statements. These JSL statements perform a Bivariate analysis of Waist versus Mass for the JMP sample data table Body Measurements.jmp.

```
body_dt = Open( "$Sample_Data\Body Measurements.jmp" );
body_biv = body_dt << Bivariate( Y( :Waist ), X( :Mass ) );

body_biv << Fit Line( {Report( 1 ), Confid Shaded Fit( 1 )});
body_biv << Fit Polynomial(2) ;

//Ask JMP to expose the locations via the report command
rpt_biv =  body_biv  << report;

body_biv << Show Tree Structure;
```

Figure 6.9 Display Box Object Path Syntax

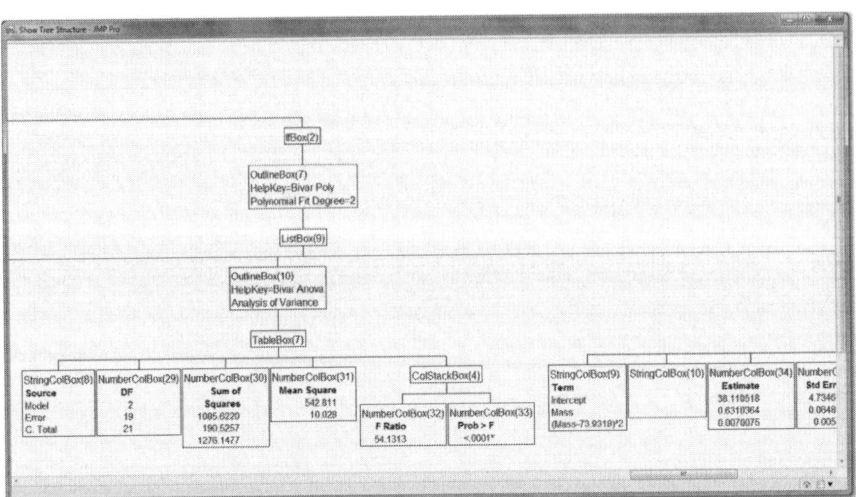

First, a few JMP report navigation rules need to be detailed. A navigation rule is referenced by its number in subsequent sections.

1. A display box is referenced by rpt[Class Name(#)]. If a display box object (dbo) has a class name assigned to it, it can be referenced by its name, rpt[Class Name ("dbo_name")]. If the dbo_name is not unique, JMP selects the dbo_name with the lowest number (#). Class names are OutlineBox, TableBox, NumberColBox, AxisBox, etc.

2. rpt[OutlineBox("name")] is equivalent to rpt["name"]. Currently, this convention only applies to OutlineBox. JMP refers to the name in an OutlineBox as a report search string. The search string must be complete and an exact match (for example Means and Std Deviations). However, the report search string can include a ? wildcard (for example, rpt["Means ? Std?"]. If a variable is used as the search string, the shorthand convention is not allowed, instead include the class name, rpt[OutlineBox(varstr)].

3. The ? wildcard can be used with other named display box objects. The OutlineBox, NumberColBox, and StringColBox objects are most likely named in a JMP report.

4. The complete path of a display box object is consistent with the syntax for displaying objects in a nested list. For example, if myList = { t1=a, t2 ={ x, y }, t3 = { d, {e, f} } }, then myList["t3"][2][1] returns e.

5. A variable can be assigned to any display box. Its address can then be used to reference display box objects that belong to it. This can simplify the syntax for deep nesting, which can result in long paths.

Run the block of code identified by the expression **hilite**. Here is a code snippet:

```
hilite = Expr(::xx << select; wait(2); ::xx << deselect; Wait(2) );
xx = rpt_biv[NumberColBox(30)];
hilite;
xx = rpt_biv[TableBox(7)][NumberColBox(2)];
hilite;
```

The **Polynomial Fit Degree=2**, **Analysis of Variance**, **Sum of Squares**, and **NumberColBox** are highlighted and unhighlighted. This demonstrates the equivalency of the following paths:

```
rpt_biv[NumberColBox(30)]
rpt_biv[TableBox(7)][NumberColBox(2)]
rpt_biv[OutlineBox(10)][TableBox(1)][NumberColBox(2)]
rpt_biv[OutlineBox(7)][OutlineBox(4)][TableBox(1)][NumberColBox(2)]
rpt_biv[IfBox(2)][OutlineBox(1)][OutlineBox(4)][TableBox(1)]
  [NumberColBox(2)]
rpt_biv[OutlineBox(1)][IfBox(2)][OutlineBox(4)][TableBox(1)]
   [NumberColBox(2)]
```

The first path is an absolute reference for **NumberColBox**. Each subsequent reference is a relative path. The following syntax demonstrates navigation rule #5: assigning a reference to a display box can simplify the syntax for deep nesting. Using references and wildcards can make navigation easier and less susceptible to version layout changes. Also, by knowing that a polynomial fit of degree 2 was scripted, and knowing the typical report layout for each fit in the Bivariate platform, this path could have been found without displaying its tree!

```
rpt_ply = rpt_biv["Poly?"]["Analy?Var?"];
xx = rpt_ply[NumberColBox("Mean?Squa?")];  // for a change
```

Absolute Versus Relative Referencing

You are probably familiar with the concept of absolute versus relative references. Maps and addresses are everyday examples of absolute versus relative references. A global positioning system (GPS) can provide the absolute reference of a target location in latitude and longitude coordinates that do not change as you continue to travel. Most GPS products can also provide directions to the target location in relative units.

The JMP tree structure provides absolute references for the display box objects in a JMP report. When the **Show Tree Structure** command is sent to a display box object (e.g., dbo_A), the window displays a map of all subordinate structures, labeled with the global coordinates (references). It is important to keep in mind that the numbers displayed in a tree structure are not the relative coordinates of the display

However, the script can set a reference to the current location (for example, display box object dbo_A). It can then specify the path in relative terms. This eliminates the need to look at the global tree or track the total number of each display box in the report.

The following example provides even more incentive for using relative references.

Relative referencing is an important concept for anyone who uses spreadsheet formulas. For example, consider an Excel spreadsheet with a formula in cell Z1 (column Z row 1) = W1*A1. If that formula is copied to cell Z2, its formula is W2*A2. Excel copied that formula with relative references, which is a wonderful convenience. If the formula in Z1 was W1*A1, then the formula in Z2 would be W2*A1. A1 is an absolute reference. A relative spreadsheet formula offers no advantage to cell Z1. Its value is realized when copying the reference to another cell.

This spreadsheet example is analogous to methods we recommend for navigating JMP reports. A single analysis can be repeated numerous times for different Y, X, and By variables. The structure of the single analysis can be applied to many analyses once a reference is assigned to the top of each analysis. I might be customizing the fifteenth instance of a Bivariate, but by starting at the top of the fifteenth bivariate, I can use the same path that I used in the first Bivariate.

This feature can be considered a navigation rule, and we can add it to our existing list.

6. Assign a reference to the top of each unique analysis (that is, a unique combination of roles), and then apply the references of a single analysis.

Single Analysis Structure and Examples

In this chapter, a single analysis corresponds to a single instance of a platform. For example, for the Oneway and Bivariate platform, there is one X, one Y, and no By columns (1X/1Y/0By). For the Distribution platform, there is one Y and no By columns (1Y/0By). For the Fit Model platform, there is one Y, one Model, and no By columns (1Y/1Model/0By). And, for the Variability platform, there is one Y, multiple X, multiple grouping, and no By columns (1Y/mX/0By).

The next example demonstrates the Bivariate and Distribution navigation and display box editing for the single analysis. This technique is later extended to multiple analyses using almost identical syntax. The example begins at Block #3 of Code in the 6_Navigation_dboPathSyntax.jsl script.

```
//----- Step #3a Editing the Analysis Layer -------------------------
//two platforms with different structure
A_body_biv = body_dt << Bivariate( Y( :Chest ), X( :Mass ) );
B_body_dist = body_dt << Distribution( Y( :Chest ), Horizontal(0) );

//--- Add customizations that are global for the report and that can
//    be made at the top layer here.
A_body_biv << Fit Line( {Report( 1 ), Confid Shaded Fit( 1 )} );
```

```
B_body_dist << {Normal Quantile Plot( 1 ),
       Fill Color( "Medium Dark Purple" ), Histogram Color( 24 )};
//---- end of step #3a -----------------------------------------
```

The following block of code labeled Step #3b customizes a Bivariate report. It is similar to the previous section's example. It is provided here as a basis for multiple variable reports. So, if this section has been difficult for you, familiarity might be refreshing at this point!

```
//---- Step #3b Customizing the Report Layer - Bivariate example ---
//---- almost identical code will be used for Multiple Ys in
//     Script 6_MultipleVariable Structure
A_body_biv << Bring Window To Front; wait(1);
A_body_biv << Fit Polynomial( 3 ); wait(2);
A_body_biv << {Curve[2] << Report( 0 ) << Line width( 3 )
    << Line Color( "Dark Gray" )}; wait(2);
rpt_A = A_body_biv << report;
rpt_A[AxisBox( 1 )][Text Edit Box( 1 )] << Set Font Size( 11 );
rpt_A[AxisBox( 1 )][Text Edit Box( 1 )] << Set Font Style( "Bold" );
rpt_A[AxisBox( 2 )][Text Edit Box( 1 )] << Set Font Size( 11 );
rpt_A[AxisBox( 2 )][Text Edit Box( 1 )] << Set Font Style( "Bold" );
//---- end of step #3b -----------------------------------------
```

Now, let's customize the Distribution report.

```
//---- Step #3c Customizing the Report Layer -Distribution example --
//---- almost identical code will be used for Multiple Ys in
//     Script 6_MultipleVariable Structure
B_body_dist << Bring Window To Front;
rpt_B = B_body_dist << Report;
rpt_B["Chest"][AxisBox( 2 )] << select;
rpt_B[Outline Box( "Chest" )][AxisBox( 2 )] << deselect;
rpt_B["Chest"][AxisBox( 1 )] << Add Axis Label( "Chest" );
rpt_B["Chest"][AxisBox( 1 )][Text Edit Box( 1 )]
 << {Set Font Size( 11 ), Set Font Style( "Bold" ), Rotate Text
(Horizontal)};
//---- end of step #3c -----------------------------------------
```

A Few Items to Note

- Axis labels are typically TextBox or TextEditBox. A TextEditBox enables the user to double-click and edit the text. A TextBox does not. JSL can edit the text for both. However, the address must use the correct class name.
- Both the Bivariate and Oneway platforms have the same messages for the same class of display objects. The Bivariate example sent one message at a time. The Distribution example bundled the edits.
- The histogram for the Distribution platform does not have a label. One was added with the message Add Axis Label.

The following block of code includes bonus commands to produce the display in Figure 6.10. Because JMP uses absolute references, it might be unnecessary to regenerate the report layer when adding a new feature like the CDF Plot. However, some displays change with options. Changing the orientation of a Distribution display has an effect on the AxisBox. We recommend regenerating the report. The example customizes the title in the outline box, so its search string has changed. A simple adjustment—adding a wildcard—makes it easy to reference subordinate display box objects.

```
//----Step #3d You can add more features to the analyses - we
recommend
//----- regenerating the report
B_body_dist << Bring Window To Front; wait(1);
B_body_dist << CDF Plot(1);
stats = rpt_B["Moments"][TableBox(1)] << Get As Matrix;
B_body_dist << {Quantiles(0), Moments(0)};
rpt_B["Chest"] << Set Title ("Chest" || " Mean = "
      || Char(stats[1], 5,2)
      || "  StdDev = " || Char(stats[2], 5,2) );
//--Note if you change the outline box search string you must adjust
//   the search string
rpt_B = B_body_dist << Report;
rpt_cdf =  rpt_B["Chest?"]["CDF?"] [PictureBox(1)];
```

Figure 6.10 Single Distribution Analysis–Navigation and Messages

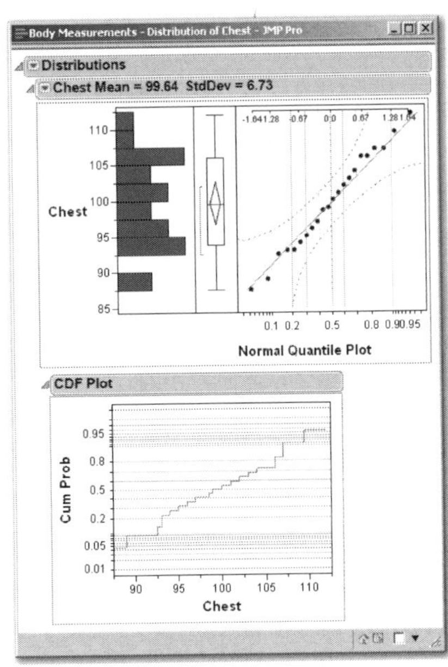

Multiple Variable Report Structure

There is one question to answer to be able to deploy navigation rule #6. How do you find the top of each analysis when a report has multiple analyses? The same JMP Sample Data table Body Measurements.jmp is used in this section. The same JMP platforms Bivariate and Distribution are used to study the structure of multiple analyses. And, once this information is mastered, the effect of By variables is discussed.

Multiple Ys and Xs and No By or Grouping Columns

The JSL examples in this section use the 6_MultipleVariableReportStructure.jsl script. It is a multiple Y variable version of the 6_Navigation_dboPathSyntax.jsl script from the previous section. Five column names are used for Y instead of one Y column name. That is the only difference in the JSL statements to generate the analysis.

```
//-----------------Generate Analysis----------------------------------
::yList = {"Fore", "Bicep", "Chest", "Waist", "Thigh"};
A_body_biv = body_dt << Bivariate( Y( eval(yList) ), X( :Mass ) );
B_body_dist = body_dt << Distribution( Y( eval(yList) ),
      Horizontal Layout(0) );

//---------------Discussion Points-------------------------------------
Show( A_body_biv, B_body_dist );
Show( Type( A_body_biv ), Type( B_body_dist ) );
Show( A_body_biv << class name, B_body_dist << Class Name );
```

Figure 6.11 Multiple Y Variable Analysis Results

```
A_body_biv = {Bivariate[], Bivariate[], Bivariate[], Bivariate[], Bivariate[]};
B_body_dist = Distribution[];
Type(A_body_biv) = "List";
Type(B_body_dist) = "Scriptable";
A_body_biv << class name = {"OutlineBox", "OutlineBox", "OutlineBox", "OutlineBox", "OutlineBox"};
B_body_dist << Class Name = "OutlineBox";
```

In the previous section, the reference variable assigned to the Analysis Layer was a single Bivariate analysis and one Distribution analysis. Figure 6.11 shows that the reference to the multiple Y Bivariate analysis is a list of five Bivariate analyses. However, there is only one Distribution analysis for all variables when clearly the display shows five distinct Distribution analyses.

Review and run the next set of JSL statements:

```
//  Platform (object) messages are items listed on the pulldowns.
//  For both A and B, if they are sent to main object
A_body_biv << Fit Line( {Report( 0 ), Confid Shaded Fit( 1 )} );

B_body_dist << {Normal Quantile Plot( 1 ),
  Fill Color( "Medium Dark Purple" ), Histogram Color( 24 )};

//------------Show syntax for local analysis options----------------
```

```
A_body_biv[1] << (Curve[1] << Line Color( "Blue" ));    //Fore Blue
A_body_biv[3] << (Curve[1] << Line Color( "Orange" )); //Chest Orange

B_body_dist[2] << Shadowgram( 1 );          //Bicep Shadowgram
B_body_dist["Chest"] << Shadowgram( 1 )
    << Histogram Color( "Blue" );           //Chest Blue Shadowgram
```

Global commands and local customizations that are set from the platform pull-down menus can be sent directly to the reference for the multiple analyses. Individual Distribution analyses can be accessed by name or by number. Individual Bivariate analyses must be accessed by number. The number is the analysis order. This approach is documented as navigation rule #7.

Recall that to gain access to the display box objects, you need to request JMP to expose the report structure and then assign it to a variable.

```
rpt_biv = report( A_body_biv[3] );
rpt_biv << Show Tree Structure;
```

This script generates a tree structure that displays the Bivariate structure for Chest versus Mass with two fitted curves, a line fit and a 2nd order polynomial fit. Notice that the addresses of the enabled features in the tree are global, exactly as expected.

```
rpt_dist = report( B_body_dist[3] );
rpt_dist << Show Tree Structure;
```

The last two commands generate the tree structure for all five Distribution analyses. The nesting structure of the two platforms is different, so the syntax to navigate the two platforms must be different. Don't despair! The fix is easy, and all JMP platforms seem to align with one of the two platforms. Here is the JSL syntax to get to the top of each analysis:

```
Report( A_body_biv[3] ) << Show Tree Structure;  //Listed

Report( B_body_dist ) ["Chest"] << Show Tree Structure; //Unlisted
```

We now have navigation rule #7, and we can add it to our existing list.

7. The reference to the top of each unique Y analysis is platform dependent. If it has the A format with Listed output, reference each analysis by its number, and then generate the report. If it has the B format with Unlisted output, generate the report for the entire analysis, and then reference the analysis by the Y column name.

The following statements create the report B_body_dist. Each Distribution analysis is referenced by name.

```
rpt_dist = Report( B_body_dist );
For( i = 1, i <= N Items( ::yList ), i++,
   rpt_dist[Outline Box( ::yList[i] )][PictureBox( 4 )][TextBox(1)]
    << {Set Font Style( "Bold" ), set Font Size(11)};

   rpt_dist[Outline Box( ::yList[i] )][AxisBox( 1 )]
    [Text Edit Box( 1 )]
     << {Set Font Size( 11 ), Set Font Style( "Bold" ),
     Rotate Text (Horizontal)};
);
```

Figure 6.12 Multiple Variable Results

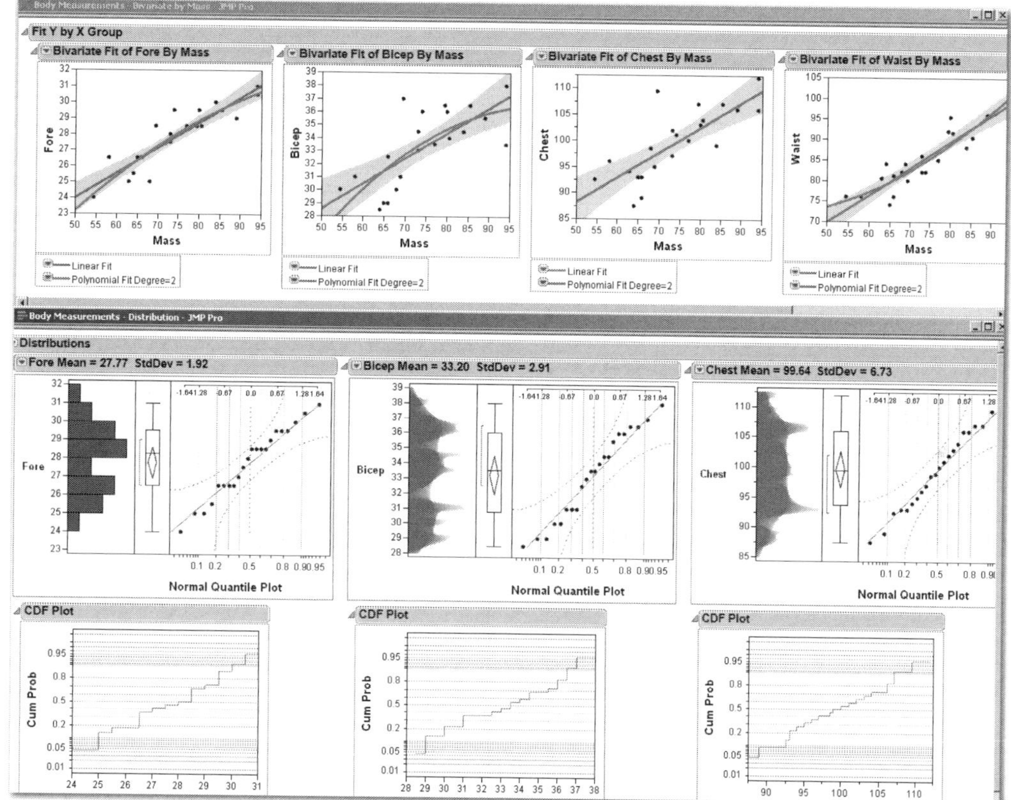

Navigate a Report with By Variables

When a platform command contains a **By( )** statement, JMP creates a grouping outline for each unique combination of By values. This adds another layer to the JMP report. After you run the following script, look at both the Bivariate and Distribution report windows, look at the Log window, and then study the following graphic.

```
//-----------------Generate Analysis----------------------------

::yList = {"Fore", "Bicep", "Chest", "Waist", "Thigh"};

A_body_biv = body_dt << Bivariate( Y( eval(yList) ),
   X( :Mass ), By(:Group));

B_body_dist = body_dt << Distribution( Y( eval(yList) ),
   By(:Group), Horizontal(0) );
```

Figure 6.13 Five Y Analyses and One By Group (S,M, B): (L) Outlines and (R) Log Window

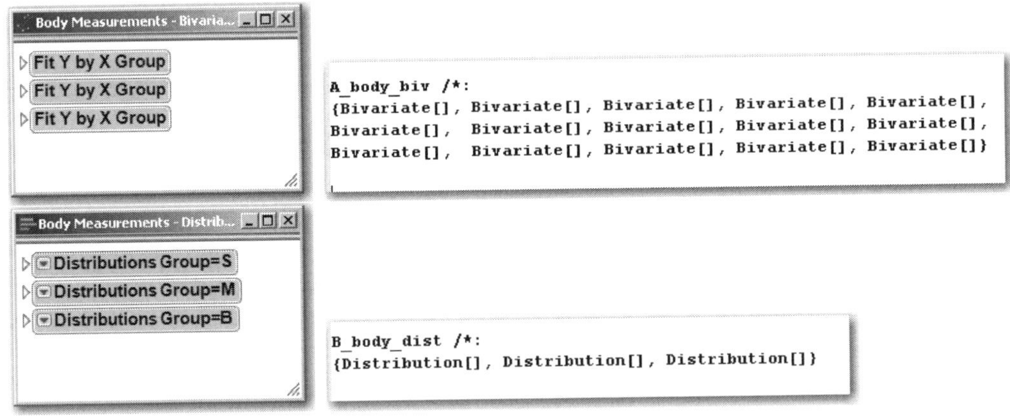

Figure 6.13 displays the closed outlines and Log window for the two JSL platform commands. The reference assigned to A_body_biv is a list of 15 Bivariate analyses (five Y*1 By Group values with three unique group values). If there is no need to modify the grouping outlines, you can apply the same methods from the previous sections to each of the 15 Bivariate analyses using a single For loop (for 1 to 15), or a nested For loop (for each By Group value and for each Y value).

The reference assigned to B_body_dist is a list, with one item for each grouping outline. Each item in the list, B_body_dist[i], is a distribution for five unlisted Y's. The solution is to access each Distribution analysis and apply the methods from the previous sections.

Figure 6.14 is a graphic that might help you picture the relationships with multiple analyses (Ys and Xs) and multiple By Groups for two different types of platform output.

A Format: Fit Y by X, Bivariate, Oneway, Logistic, Contingency, Variability, ContourPlot

B Format: Distribution, MatchedPairs, Fit Model, Control Chart

Figure 6.14 (L) A Format with Listed Output (R) B Format with Hidden WrapList

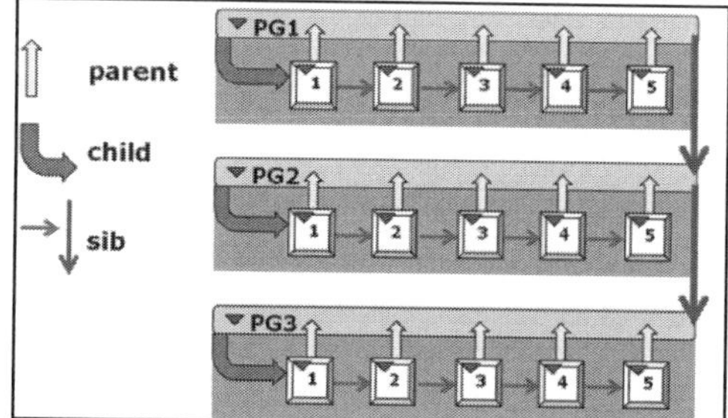

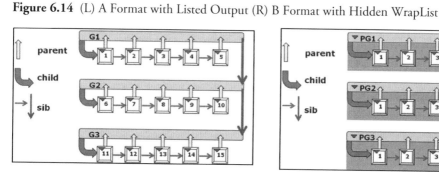

A Few Items to Note

- Each platform has its own layout of display box objects. You should learn how to reference items for one analysis.
- Navigation paths with search strings and wildcards might not need to change when platforms are upgraded with new features and layout changes.
- For multiple-variable navigation, create a reference to the top of each analysis, and then apply the paths of a single analysis.
- The 6_NavigateReportwithByVariables.jsl script provides examples of accessing and modifying the grouping outlines. It also provides examples of parent-child-sib relationships and how to maintain the same scale across all By Groups (a common request when creating By analyses).
- This section tackled the task of editing multiple JMP analyses. Generating multiple analyses is a simple task when you have lists of Y and X and By variables, but it requires a little extra skill to customize display box objects using the top-down approach presented in this chapter. Chapter 8 includes a section on building a display bottom up. In this type of display, each analysis is generated and customized one at a time, and then it is added to a user-defined display layout.
- Find( ) and Xpath( ) are two JMP functions that can be useful for locating objects in a display. Examples and documentation can be found in the JMP online Help by selecting **DisplayBox Scripting Index ► Wraplist**. Xpath( ) is new in JMP 9, and has huge potential for customizing complex displays with less navigation.

Extract Information from a Report

JMP stores its analysis results and summary statistics in the report window's TextBox and TableBox objects. **Make Into Data Table** and **Make Combined Data Table** are two useful messages for extracting data stored in report TableBox objects. **Make Combined Data Table** creates a single table for the multiple analyses belonging to the same Report Layer.

The following example is a simulation of modeling the behavior of a processing tool in a semiconductor factory. Temperature readings are taken at five different times while a wafer is being processed. Three wafers are sampled from each batch of 25 wafers. A batch of wafers is called a "lot." The last 64 lots of raw data representing readings from four different processing tools are stored in a JMP table. The rate of rise, its linearity (or lack thereof), and the time it takes to reach a target temperature of 240 C are indications of the state of a processing tool. The states are stable, unstable, on target, off target, etc.

The analysis creates 64 Bivariate analyses, each with three curves. Wfr is a By Group column. Each Bivariate analysis represents a unique Tool and Lot combination. Each of three sampled wafer's values within a lot gets a separate quadratic fit, for a total of 192 regression summaries. Just imagine performing these analyses using the point-and-click method!

Make Combined Data Table

The **Make Combined Data Table** command is incredibly powerful. It can be applied to any TableBox. In this example **Make Combined Data Table** is applied to the Summary of Fit TableBox to create a data table that contains information for however many analyses are in the report (in this example, 192). Figure 6.15 (L) displays a portion of the combined table of the Summary of Fit.

Typically, these simple commands are just the beginning steps to produce the appropriate output. In this instance, the appropriate output is a table of model results, where each row (192 rows) is a regression summary that includes identifiers for Lot, Tool, Wfr, parameter estimates, and fit diagnostics. Scripting work needs to be performed to get the table into the correct format.

```
ror_stk_dt = open("$JSL_Companion\ROR Model Format.jmp");

ror_biv = ror_stk_dt << Bivariate(Y(:Temperature), X(:Time),
    By(:Tool, :Lot), Group By(:Wfr), Fit Polynomial(2) );

tmp=Report( ror_biv[1] )[Outline Box( 1 )]["Summary of Fit"]
       [Table Box( 1 )] << Make Combined Data Table;
```

Figure 6.15 Summary Of Fit: (L) Make Combined Data Table (R) Split Table

	Lot	Tool	Wfr	Column 1	Column 2
1	L01115196	ent01	1	RSquare	0.9999599636
2	L01115196	ent01	1	RSquare Adj	0.9999199273
3	L01115196	ent01	1	Root Mean Square Error	0.7869507094
4	L01115196	ent01	1	Mean of Response	130.36
5	L01115196	ent01	1	Observations (or Sum Wgts)	5
6	L01115196	ent01	2	RSquare	0.9986506135
7	L01115196	ent01	2	RSquare Adj	0.9973012269
8	L01115196	ent01	2	Root Mean Square Error	4.5573014677
9	L01115196	ent01	2	Mean of Response	131.26
10	L01115196	ent01	2	Observations (or Sum Wgts)	5

	Tool	Lot	Wfr	Root Mean Square Error	RSquare	RSquare Adj
1	ent01	L01115196	1	0.7869507094	0.9999599636	0.9999199273
2	ent01	L01115196	2	4.5573014677	0.9986506135	0.9973012269
3	ent01	L01115196	3	0.9294312496	0.9999430396	0.9998860791
4	ent01	L01115198	1	3.6698471491	0.9990451827	0.9980903654
5	ent01	L01115198	2	3.779673824	0.9989791593	0.9979583187
6	ent01	L01115198	3	2.7353605557	0.9994690236	0.9989380472
7	ent01	L01115200	1	1.7858820025	0.9997714854	0.9995429708
8	ent01	L01115200	2	2.2171602395	0.9996447597	0.9992895193
9	ent01	L01115200	3	1.5168104627	0.9998334647	0.9996669294

Splitting Column 2 of the tmp table by Column 1 returns a table of Summary of Fit information in the correct format. This takes just a couple of JSL statements (not all columns are displayed). (See Figure 6.15 (R)).

```
// split tmp
// where each fitted model is a row/obs and assign a reference
fit_sum = tmp << tmp << Split(
   Split By( :Column 1 ),
   Split( :Column 2 ),
   Group( :Tool, :Lot, :Wfr ),
   Remaining Columns( Drop All )
);
Wait( 0 );
Close( tmp, NoSave );
```

The same techniques can be applied to Parameter Estimates, and then the two tables can be joined. Well, maybe that would work, if the regression fit was an uncentered fit. Figure 6.16 is the table returned by the **Make Combined Data Table** command for Parameter Estimates with the centered polynomial fit. It is sorted so that the problem of splitting by Term is visible. The centered polynomial fit is the JMP default, which occurred because no option was specified. Splitting column Estimate by column Term does not return the appropriate results.

Figure 6.16 Parameter Estimates (Centered) Returned by Make Combined Data Table Command

8/0 Cols	Lot	Tool	Wfr	Term	Estimate	Std Error	t Ratio	Prob>|t|
1	L01115191	ent32	1	Intercept	26.980864535	1.0804353042	24.97	0.0016
2	L01115191	ent32	1	Time	10.593049871	0.0739143602	143.32	<.0001
3	L01115191	ent32	1	(Time-10.0067)^2	-0.023606123	0.0121098026	-1.95	0.1906
4	L01115193	ent21	1	Intercept	22.260813468	0.815467588	27.30	0.0013
5	L01115193	ent21	1	Time	10.879470959	0.0547922741	198.56	<.0001
6	L01115193	ent21	1	(Time-9.96667)^2	0.0085090975	0.0093308036	0.91	0.4581
7	L01115196	ent01	1	Intercept	23.187028583	0.7047705066	32.90	0.0009
8	L01115196	ent01	1	Time	10.939030189	0.0490722995	222.92	<.0001
9	L01115196	ent01	1	(Time-9.92)^2	-0.063589032	0.0079524497	-8.00	0.0153
10	L01115198	ent01	1	Intercept	19.626365094	3.7914062374	5.18	0.0354
11	L01115198	ent01	1	Time	11.12515686	0.2454584822	45.32	0.0005
12	L01115198	ent01	1	(Time-10.0067)^2	0.0858628765	0.0441163754	1.95	0.1910
13	L01115200	ent01	1	Intercept	32.04172565	1.6294565432	19.66	0.0026
14	L01115200	ent01	1	Time	10.408738657	0.1113693467	93.46	0.0001
15	L01115200	ent01	1	(Time-10.0133)^2	-0.034511133	0.0182487686	-1.89	0.1992

Changing the analysis statement in the script to be uncentered is one option.

```
tmp = ror_biv = ror_stk_dt << Bivariate( Y( :Temperature ),
   X( :Time ), By( :Tool, :Lot ), Group By( :Wfr ),
Fit Special( Degree( 2 ), Centered Polynomial( 0 ) ) );
```

There are four scripts associated with this section.

- 6_ExtractInformation_GetTableInfo_Centered.jsl
- 6_ExtractInformation_GetTableInfo_unCentered.jsl
- 6_ExtractInformation_TableInfo_Traverse_Centered.jsl
- 6_ExtractInformation_TableInfo_Traverse_unCentered.jsl

These scripts are intended to show that there is more than one way to get the same results. Some ways are better than others. The scripts provide solutions to common issues that other scripters have faced (like getting the equations into the appropriate format and parsing strings). And, these scripts reinforce the point that analysis methods matter and there are trade-offs with different platforms.

A Few Items to Note

- The TableBox message Make Into Matrix is not discussed in this section. Make Into Matrix can be useful for getting summary information from a single table, like a Variability summary table.

- All four scripts for this section start with a single report window with the entire analysis assigned to one reference—ror_biv. The scripts, using the navigation rules for report tree structures, reference the top of each analysis, and customize and get information for the individual analyses using display box messages. We refer to this method as *top-down customization.*

- The ability to extract specific information from text strings is a valuable skill for many scripting tasks. When getting information from JMP reports, this skill is essential. The pattern matching functions in JMP include RegexMatch, which executes regular expression matching. JMP provides many functions that you might find more intuitive than the shorthand syntax of regular expressions.

  ```
  //use pattern matching to get the entity and lot values
    p1 = Pat Arb() + "Tool=" + Pat Arb() >? entid + "," + Pat
  Arb()
      + "Lot=" + PatRem()>? lotid;
    Pat Match(ttl, p1);
  ```

  In this pattern, ttl is a title from a Bivariate outline box. p1 defines a pattern. Pat Match( ) applies the pattern to the title. The syntax >? assigns values. After Pat Match( ) is executed, entid will contain all characters in the title between Tool= and the comma. lotid is everything after Lot=. The scripts for this section include alternatives using JMP character functions to get Tool and Lot information. Regular expressions can be exasperating to write and read. JMP character functions can be easier to understand both today and in the future.

- The Bivariate option for Centered Polynomial (0 | 1) is in the Fit Special message.
- Several JMP analysis platforms have options to save models, formulas, and fitted values as new columns in the source table or as a script.
- The Fit Model platform includes several JSL-only messages to get model diagnostics. See the *JMP Scripting Guide* for more details. Two common messages are the following:

  fit_model_object << Get Estimates( );

  fit_model_object << Get Parameter Names( );

Create Custom Reports

In our experience, the general progression of a JSL student is to learn the commands for their favorite platform, and then learn how to generate a By variable report. Then, the JSL student learns how to customize titles, remove portions of reports, and make other changes such as arranging plots and getting multiple platforms in the same window.

Often, a few simple commands can be added to a script to provide output that meets your needs or is close enough without building a custom display as discussed in Chapter 8.

Before proceeding, select **Help ► JSL Functions**. In the categories, find and study New Window, HListBox, VListBox, and LineupBox. It is important to recognize that a display box is invisible until it is exposed in a report window.

Recall the script from the previous section that creates 64 Bivariate analyses, each with three curves, all in a single window. Nesting the following command in a **LineupBox( )** creates a display box that allows restructuring the layout. However, the display box is invisible unless it is nested in a report window.

```
ror_biv = ror_stk_dt << Bivariate( Y( :Temperature ),
  X( :Time ), By( :Tool, :Lot ), Group By( :Wfr ),
  Fit Special( Degree( 2 ), Centered Polynomial( 0 ) ) );
```

The first command lays out the analyses four to a row. The second command changes the layout to two per row. This code uses the general rule of creating a reference to each object, which makes the objects easier to use later.

```
ror_nw = New Window("Temperature Rate of Rise",
  lub = LineupBox( ncol( 4 ),
    ror_biv = ror_stk_dt << Bivariate( Y( :Temperature ), X( :Time ),
      By( :Tool, :Lot ), Group By( :Wfr ),
      Fit Special( Degree( 2 ), Centered Polynomial( 0 ) )  )
)); //end lineupbox and new window
lub << ncol(2); //lineup box may be modified later
```

Invariably, the next question is, "How do I lay out a display that partitions for a specific variable, such as tool in this example?" The code depends on the output requirements. Often, a single window containing all tools with marked partitions is wanted. How to script this depends on whether the values for tool are known or whether they are dynamic. This section addresses the scenario where the tool values are known. For the other scenario (where the values are dynamic), we suggest the coding style in Chapter 8. Recall that there are four tools for this analysis—ent01, ent21, ent32, and ent44. This first step adds a **Where( )** condition to the previous Bivariate code, wraps it in a **LineupBox** and an **Outline Box**, and repeats this four times.

```
ror_Tool_nw = New Window("Temperature Rate of Rise --- By Tool
Layout",
  ob1 = Outline Box(toolList[1],
    lub1 = LineupBox(ncol(4),
      ror_biv1 = ror_stk_dt << Bivariate(Y(:Temperature), X(:Time),
        By(:Lot), Where( :Tool == toolList[1] ),  Group By(:Wfr),
        Fit Special( Degree( 2 ), Centered Polynomial( 0 ) )  )
   )), //end ob1, note a comma separates ojects in New Window
//< Not shown... ob2 and ob3 that have the same patterns as ob1, ob4>
  )), //end ob3, note a comma separates ojects in New Window
  ob4 = Outline Box( toolList[4],
    lub4 = LineupBox( ncol(4),
      ror_biv4 = ror_stk_dt << Bivariate(Y(:Temperature), X(:Time),
      By( :Lot ), Where( :Tool == toolList[4] ), Group By( :Wfr ),
        Fit Special( Degree( 2 ), Centered Polynomial( 0 ) )  )
  )) //end ob4, no comma end of display boxes
);    // end of new window
```

The 6_CustomReports_FixedPartitions.jsl script contains the code from the previous examples. It includes commands to clean up the Bivariate titles, turn off the quadratic fit reports, and close the outline box for each tool. The focus is on one tool at a time. This script demonstrates that nesting a few layout commands can dramatically change the report.

Figure 6.17 Multiple Analyses Partitioned by Tool Using Outline Box and LineupBox

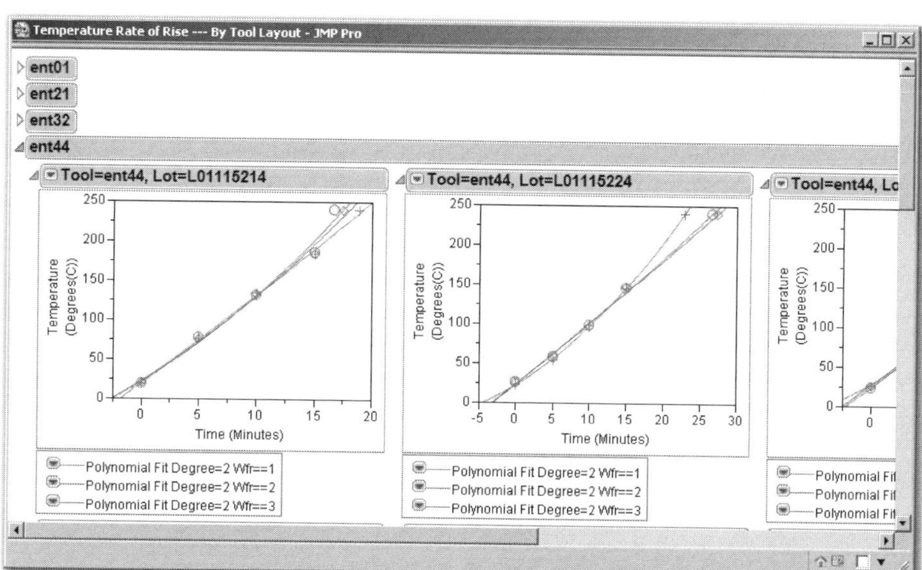

The previous section provided two methods for extracting the regression estimates and Summary of Fit values into a table. It stated that tool performance is reflected in the stability of rate of rise. The rate of rise involves the start temperature (t0), end point time (ep), the time the tool reached 240 C, and Therms (thm), a calculated metric of total temperature time.

The 6_CustomReport_Trends.jsl script uses the rate of rise data and provides JSL commands to create comparative trend plots of summary data. It adds a special report to each tool-lot graphic, and uses Graph Builder to create trend plots that are prepended to the original report. There is one preparation step to make this easier—add an Outline Box above the LineupBox. The outline box tlp_ob1 becomes the owner or parent of the LineupBox that arranges the tool outlines. This allows the separation of the tool lot reports from the overall trend plots of all tools and lots.

```
ror_nw=New Window("Temperature Rate of Rise",
  tlp_ob1 = Outline Box("Tool Lot Plots", //added to partition plots
    lub=LineupBox(ncol(4),
      ror_biv = ror_stk_dt << Bivariate(Y(:Temperature), X(:Time),
...
```

Add the JSL statements to the script to create the trend plots. Use **Sib Prepend** for the new outline box named tlp_ob1.

```
tlp_ob1 << Sib Prepend(  //Load 3 outline boxes >>before<< tlp_ob1
lub2 = LineupBox(ncol(1),
    t0_ob1 = OutlineBox("T0, Temperature (C) at Start",
          t0_HL=HListBox()),
    ep_ob1 = OutlineBox("EP, Minutes to Reach 240 C",
          ep_HL=HListBox()),
   thm_ob1=OutlineBox("Therms, Total Temperature/Time",
       thm_HL=HListBox())
));
//By wrapping this in a VListbox the Graph Builder report window
//is still live.
thm_vv = VListBox(thm_gb=rpt_dt<<Graph Builder(), ...
thm_HL<< Append( thm_vv ); //Load >>into<< thm_HL
```

Figure 6.18 Temperature Rate of Rise Trend Plot

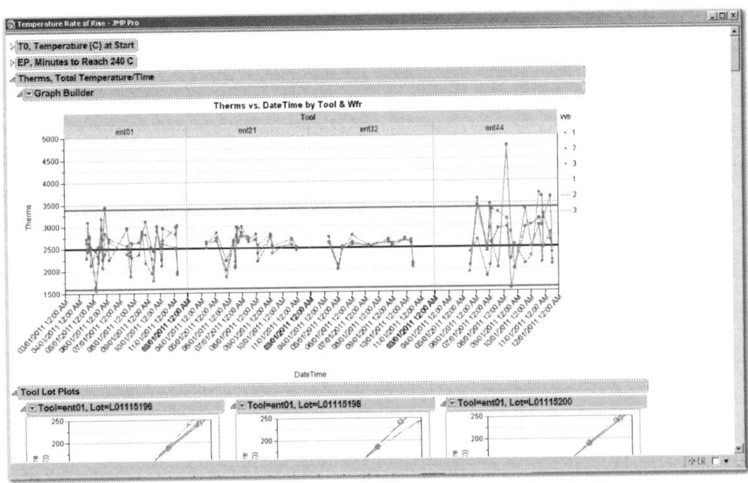

Figure 6.19 Tool Lot Bivariate with Custom Report and Highlighting

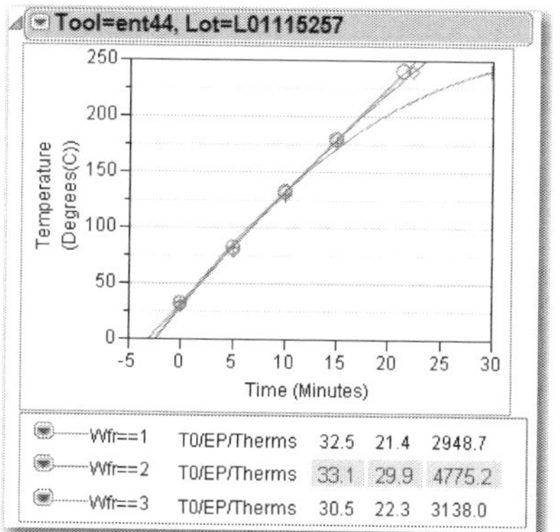

The report window in Figure 6.18 is the final result of many of this chapter's learning objectives. The report window effectively addresses the stated goal of the example. By mastering a few commands, this report includes live graphs from two different tables and other special features. Figure 6.19 displays one of these special features: the addition of important indicators with unusual values highlighted in the raw data and Summary of Fit table. Reports like these can make decisions easier by creating actionable information out of data.

A Few Items to Note

- The examples demonstrate that the simple addition of an outline box (LineupBox) can make dramatic changes in the report window.
- VListBox, HListBox, and DisplayBox are invisible until they are displayed in a report window.
- The examples in this section intentionally use explicit, hardcoded commands instead of functions and expressions. Linear explicit coding is more intuitive for first-time scripters, who often find the level of abstraction with expressions and functions confusing.
- The numerous Graph Builder options in the 6_CustomReport_Trends.jsl script were generated by JMP after a point-and-click session for the first parameter, thm. Then, the commands were copied, pasted, and edited for the other two parameters, ep and t0. Alternatives to cookie-cutter methods are discussed in Chapter 7, Chapter 8, and Chapter 9, "Writing Flexible Code."
- Deleting display objects is a matter of referencing and sending a delete command. Extra caution should be taken when modifying and deleting display box components from live report windows. Many scripters have asked, "How do I remove the border box that contains the curve report options?" If the border box resides in a journal window or in a window that

is not active, it is safe to delete it. If it is deleted in a live window, and an additional curve request is made from the Bivariate pull-down menu, it will cause problems.

Save Results

Once you create an analysis and have the appropriate output, the next step is to save the information. A report window or portions of a report can be saved as a JMP journal or in formats other than JMP. A journal file allows subsequent analyses to be appended. The format of the output changes when it is saved as something other than a journal file.

Experiment with the different options to determine the format that meets your needs. As you experiment, keep in mind that some features of JMP are available with a journal file, such as decorating graphs and opening and closing outlines. These features are not available with other output formats.

Save an Analysis

JMP provides many formats for saving the results of an analysis. The formats used most frequently are included in this section. Other formats can be found in the *JMP Scripting Guide*. The examples are included in the 6_SaveResults.jsl script.

```
//Define path name in which to save file
save_path = "c:\temp\";

bigclass_dt = Open( "$SAMPLE_DATA\Big Class.jmp" );

//Single analysis and append a second
bigclass_onew = bigclass_dt << Oneway(Y( :height ),X( :age ),
Means( 1 ),Mean Diamonds( 0 ),Plot Quantile by Actual( 0 ),
Line of Fit( 0 ), Box Plots( 1 ), ANOVA( 0 ),
X Axis proportional( 0 ), Points Jittered( 1 ));
//Create report
onew_rpt = bigclass_onew << Report;

//Create another analysis and append it to the first report
bigclass_biv = Bivariate( Y( :height ), X( :weight ) );
biv_rpt = bigclass_biv << Report;
onew_rpt << Append( biv_rpt );

//Save appended report in different formats
onew_rpt << Save PDF( save_path || "bigclass_out.pdf" );
onew_rpt << Save Journal( save_path || "bigclass_out.jrn" );
onew_rpt << Save MSWord( save_path || "bigclass_out.doc" );
onew_rpt << Save HTML( save_path || "bigclass_out.htm" );
```

```
// Save a specific part of the output to a text file
// Note that this can be saved even if not part of the display
onew_rpt["Means and Std Deviations"] << save Text(save_path ||
"mn_std.txt");
```

Save with By Group

Using the previous syntax, only the first item in a By Group report is saved. In this case, only the output for the males is saved. To save all objects in the By Group, a different syntax is needed. The JSL statements for saving the Bivariate analysis of **weight** versus **height** by **sex** follows. The first display box is used, and the parent of that box is found. Then, the parent of the report is found, and the entire report is saved.

```
//Saving an analysis with a By Group (see page 217 JSG)
bigclass_by_biv = bigclass_dt << Bivariate( Y( :weight ), X( :height
), By( :sex ) );
biv_by_rpt = bigclass_by_biv << Report;

((biv_by_rpt[1] << parent) << parent) << Save PDF( save_path ||
"bigclass_by.pdf" );
```

The previous section demonstrated several layout strategies that might be useful here. Using exactly the same strategy on a much simpler example, wrap a **LineupBox** or an outline box, and then journal or save to PDF.

```
//Using a LineupBox or OutlineBox container to be saved
nw=New Window("BigClass",
ob1 = OutlineBox("Gender Comparison",
lub1 = LineupBox(ncol(2),
     bigclass_by_biv = bigclass_dt
     << Bivariate( Y( :weight ), X( :height ), By( :sex ) )))
);
ob1  << Save PDF( save_path || "ob_bigclass_by.pdf" );
lub1 << Save PDF( save_path || "lub_bigclass_by.pdf" );
```

Save Multiple Analyses with New Window

When multiple analyses are created using the **New Window( )** function, the analyses can be saved by journaling the output, creating a reference to the journal, and then saving the files using the reference.

```
//Saving Multiple Analyses with New Window
bigclass_nw = New Window( "Big Class",
HListBox(
  bigclass_dt << Oneway( Y( :height ), X( :sex ),
     Means and Std Dev( 1 ), Plot Actual by Quantile( 1 ),
     Box Plots( 1 ), Mean Error Bars( 0 ), Std Dev Lines( 0 ),
     Points Jittered( 1 )),
  bigclass_dt << Bivariate( Y( :weight ), X( :height ), Fit Line )
));
```

```
//Create journal and define handle to journal
bigclass_nw << Journal;
bigclass_jrn = Current Journal();
bigclass_jrn << Save PDF( save_path || "bigclass_nw.pdf" );
bigclass_jrn << Save Journal( save_path || "bigclass_nw.jrn" );
bigclass_jrn << Save HTML( save_path || "bigclass_nw.htm" );
```

Save a Picture or Selection

There are times when you want to capture part of the JMP output and save it as a picture. JMP has a clean and easy-to-read format that a picture will retain. The following example shows how to save a picture from the report as a PNG file. Any selection from a report can be saved as a picture, and other formats besides PNG (JPG, GIF, EMF, etc.) are available.

```
//Save a picture or a report selection
bigclass_biv = bigclass_dt << Bivariate( Y( :weight ), X( :height ),
    Fit Line );
bigclass_rpt = bigclass_biv << report;
bigclass_rpt[PictureBox(1)] << Save Picture(save_path||"wt_vs_ht.png");
```

We have found these methods for saving output to be especially useful when generating reports and graphs and making them available on a Web site.

7 Communicating with Users

Introduction

Communicating with users belongs to the rubric of user interface (UI) design, also known as human-computer interaction (HCI). The primary goal of this chapter is to develop your knowledge of JMP functions and objects used to create UIs and learn how to apply them.

Being skilled with a language's components is the first step of UI design: knowing what is available and how to use it. The second step is the design itself. We are certainly not experts on how to design a good interface. As users, we have experienced many bad UI designs, such as annoying phone trees, hard to find options on Web sites or software dialog boxes, etc. We include recommendations for design, such as:

- Know your users. Are they experts, or do they need guidance or access to help?
- Lay out the information flow. Can all communication occur in one step? Is the information needed before a display is generated, or can some of the information be collected in the report UI?
- Consider the look and feel. Are the options logically organized and easy to find? Are the prompts and dialog boxes consistent? Is the skin layer (color, button layout, and display) consistent? Are the widgets (e.g., sliders or boxes) consistent? Are the controls (e.g., radio buttons or option lists) consistent? Are all the prompts familiar and consistent with the JMP UI?
- Think about usability and robustness. Are input errors captured and handled elegantly (e.g., allows a user to re-enter versus start over)? If the user is toggling many options or settings for multiple objects, are there options to set for all? Can default values be generalized?

There are many references on this topic. There are books and magazines, or you can search the Web for ISO 9241, which is a standard covering the ergonomics of human-computer interaction. A good UI is a combination of science and art.

Introduction to Dialog Boxes

When creating a script that requires flexibility and robustness, the user is often required to provide essential information during script execution. If a script is run by many users or on different data sets, it is likely that the data table (or at least the columns in a data table) must be specified. Also, depending on the analysis, the analyst might require a customized report based on information provided during the execution of the script. This flexibility is obtained in JMP by using dialog boxes to retrieve information from users. In this chapter, a *dialog box* is any window that is used for two-way communication between a user and the JSL script. A dialog box is not limited to displays available using the JMP **Dialog( )** function.

A dialog box can be something simple, where the user is asked to choose a directory or select a radio button. Or, it can be fairly complex, where the choices available to the user are dependent on previous choices made during the session. When writing a script to construct a dialog box, there are two important properties to consider. The first is whether the dialog box requires user interaction before the rest of the script can execute (modal versus non-modal). The second is creating an intuitive user interface. These properties are discussed in this section and throughout the other sections as more examples are developed.

Modal Versus Non-Modal Dialog Boxes

A modal dialog box requires user interaction before the script proceeds. The browse dialog box initiated by an unspecified **Open( )** statement is modal. All JMP graph and analysis platform dialog boxes are non-modal, meaning that the user has full access to the JMP UI (e.g., creating a new column, sorting a table, etc.) before responding.

When a non-modal dialog box, such as the platform dialog box created by the Distribution platform, is included in a script, the script does not wait for the user to respond. Subsequent statements execute while that dialog box is open. The following is an example of the difference between modal and non-modal. When the first set of code is run, the data table is opened, and a **Distribution** dialog box is opened because no variables are specified. The script immediately executes the **Bivariate** command and produces a report window with a scatterplot of Minimum Price versus Midrange Price. The **Distribution** dialog box does not wait for a response—the dialog box is still available to be completed by the user. In the second set of code, an Open Data File dialog box appears, and no other output is generated until a table is selected.

```
//Example of a non-modal dialog box
//Similar to other analysis platforms, Distribution is non-modal
cars_dt = Open( "$SAMPLE_DATA/Cars 1993.jmp" );
cars_dt << Distribution();
cars_dt << Bivariate(
  Y( :Name( "Minimum Price ($1000)" ) ),
  X( :Name( "Midrange Price ($1000)" ) )
);

//Open() is a modal dialog box
cars_dt = Open();
cars_dt << Distribution();
cars_dt << Bivariate(
  Y( :Name( "Minimum Price ($1000)" ) ),
  X( :Name( "Midrange Price ($1000)" ) )
);
```

Format of a JMP Dialog Box

JMP has a standard UI for its dialog boxes that is intuitive and familiar to users. For example, consider the dialog box for the Distribution platform. (See Figure 7.1.) The dialog box has a relevant title (Distribution). At the top left, there is a brief explanation of the platform objective. Just below the explanation, there is a box containing a list of the current data table's columns. In the middle, there are buttons to cast selected columns into roles. On the right, there are action buttons. The placement and order of the action buttons are always the same. Other JMP dialog boxes have a similar pattern. Because most users are familiar with the JMP UI, our advice is to use similar standards when creating a custom dialog box. Changing the order or placement of key items in a dialog box might slow user response or cause confusion.

Figure 7.1 JMP Distribution Dialog Box

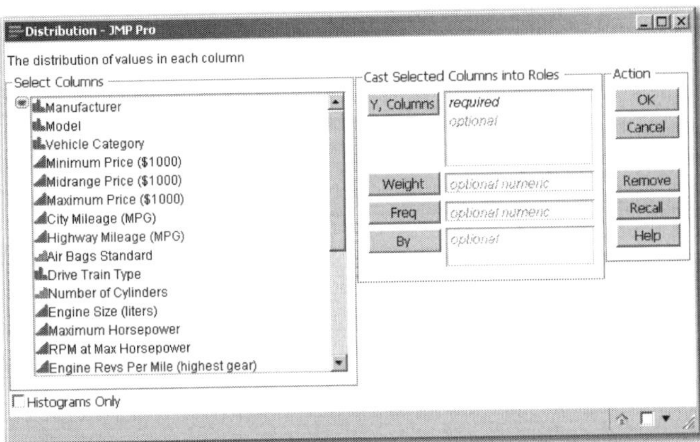

The **Distribution** dialog box is one of the simplest. If you examine other platform dialog boxes, you will see other options such as radio buttons, pop-up menus, combo boxes, etc. Many of these options are discussed in subsequent sections.

In Chapter 6, "Reports and Saving Results," we use the command Show Tree Structure( ) to create a map of display box objects in a report window. This map helps navigate to specific objects to get information and send messages to customize the report. This command is available for most dialog boxes. However, it is a little tricky to find because it is not available in a pull-down menu. To access this command, go to an open area in the dialog box (but not too close to other features such as buttons and sliders), and right-click while holding down the CTRL key and SHIFT key. If it is available for that dialog box, a **Show Tree Structure** button appears. (See the middle of Figure 7.2.) Click the **Show Tree Structure** button to display the structure and the objects used to build the dialog box.

Figure 7.2 JMP Fit Y by X Dialog Box—Show Tree Structure Button

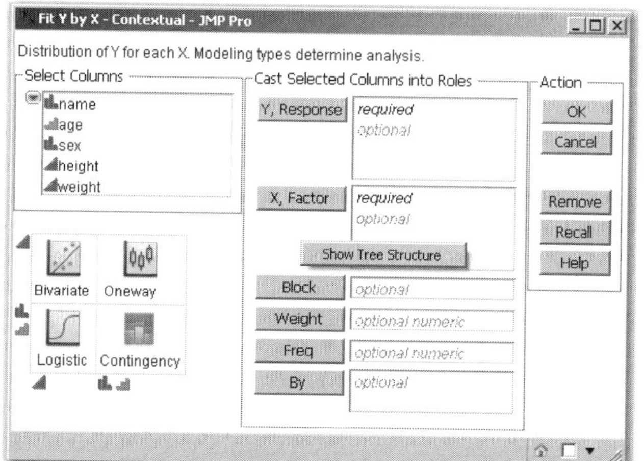

The tree structure of the **Fit Y by X** dialog box includes familiar display box objects **ListBox( )**, **ButtonBox( )**, and **TextBox( )**. JMP built-in dialog boxes might include display box objects not accessible in JSL.

As you write scripts to create dialog boxes, you will develop an artistic style. Keep in mind that the end goal is usability: the user should intuitively understand what is needed, and the user should be able to quickly reply and execute the script. Our advice is to start by emulating JMP dialog box standards. Then, as you add more features to your own dialog boxes, develop your own standards and follow them. For example, all JMP platform dialog boxes use an Action **PanelBox( )** that vertically aligns the **OK**, **Cancel**, **Remove**, **Recall**, and **Help** buttons.

Format of a Custom Dialog Box

The format of a custom dialog box is completely flexible. The same display box objects discussed in Chapter 6 for formatting a report window can be used to create a custom dialog box. For example, the simple dialog box created by the following script uses a **Lineup Box( )** with two arguments. Many of the examples in subsequent sections of this chapter include additional display box objects to lay out dialog box controls such as **VListBox( )** and **HListBox( )**.

```
New Window( "Customer Survey",
  Text Box( "Gathering Customer Information" ),
  Lineup Box(2,
    Text Box( "Name" ), Text Edit Box( "Last, First  " ),
    Text Box( "Zip Code" ), Text Edit Box( "97110" )));
```

A Few Items to Note

The 7_IntroDialogs.jsl script includes the code in this section and the following:

- A discussion of the first example. What if the user does not respond as expected?
- Another example of modal and non-modal behavior using custom dialog boxes.
- A simple example of getting the user response (his or her name) and writing a personal hello.

Specific syntax is covered in later sections of this chapter. Working through the extras included in this script sets the stage for the later sections.

Messages and Dialog Boxes

This section documents JSL utility functions that send information to the user and use the JMP **Dialog( )** function to prompt the user for information.

Messages

Users are usually interested in knowing why an expected output did not appear. Open the Sample Data table Iris.jmp, and try running the command biv = Bivariate(Y(:Sepal length), X(:Species)) ;. It does not run. Fortunately, JMP provides an error message in the Log window stating that the column Species is not numeric, and sets the value of biv to be a missing number, (.). The following message functions are found in the online Help, available by selecting **Help ▶ JSL Functions ▶ Utility**. These message functions are easy to understand and use. The 7_MessagesDialogs.jsl script demonstrates several features. We think you'll find the **Caption( )** and **Show( )** functions to be the most useful. Here's why.

- Prior to JMP 9, the status bar was in the bottom left corner of the main window, which was always resident. **Status Msg( )**, which displays a specified message written in the bottom left corner of the script, might not be visible if the script opened a table or a report window, or if the script is invisible to the user.
- **Write("text")** and **Print(expr1, expr2, ...)** have little context. **Watch( )** can be useful for debugging, but it is typically removed from production scripts. In the Log window or in a separate Script Window, type Watch(x,y), and run it. If it is important for the user to see the status of specific variables, we recommend you create a custom window with explicit contextual information and values.

- **Beep( )** might be useful to get a user's attention. **Speak( )** provides a computer voice. The **Web( )** function can initiate a human voice message, a video link, and an extensive help page. The **Web( )** function can produce a friendlier interface.

JMP Built-in Dialog Boxes

With each new version, JMP has provided more display objects and utility functions to customize a dialog box. JMP has deprecated the use of **Dialog( )**, and it might not be available in future versions. Nonetheless, **Dialog( )** is still available in JMP 9. If you are writing something quick and simple, this function might save you a lot of time. Studying the output from this function will provide more insight into the display boxes described in the next few sections. Both **Dialog( )** and **Column Dialog( )** are JMP functions that build a modal dialog box awaiting the user's response. Subsequent commands in the script do not run until the user has responded.

Pick File( ), **Pick Directory( )**, and **Open( )** were introduced in Chapter 2, "Reading and Saving Data." They are mentioned here, and examples are included in the script for this section because they are examples of modal, built-in dialog boxes. The following script prompts the user for the information that is needed to build a database query. ndays is the number of days of data to extract, param is the name or string of the parameter to extract, and step_id is a unique, 8-digit number identifying the process step.

```
//Run several times with different values or Cancel. See Log window
//or hover over usr_info for results. Note you can enter a string for
//a number and vice versa. There's no built-in error checking; it's
//the script writer's job
usr_info = Dialog("Select the query filters ",
    HList( Lineup(2,
         "Number of days, 30 day maximum", ndays = edit number(7),
         "Parameter name, wildcard=* ", param = edit text(),
         "Manufacturing Step Id", step_id = edit number()
        ),    //end of lineup
    Lineup(1, Button ("OK"), Button ("Cancel")
//Dialog button alignment ignored by JMP since v8
    ) //end of button align
)); //end of Dialog
show(usr_info, usr_info["ndays"]); //*:usr_info = {ndays = 7, param =
"NLIN1",
//   step_id = 4543, Button(1)};
```

Figure 7.3 Database Query Filter Dialog Box Using Edit Text Constructor: (L) Dialog Box (R) Log Window

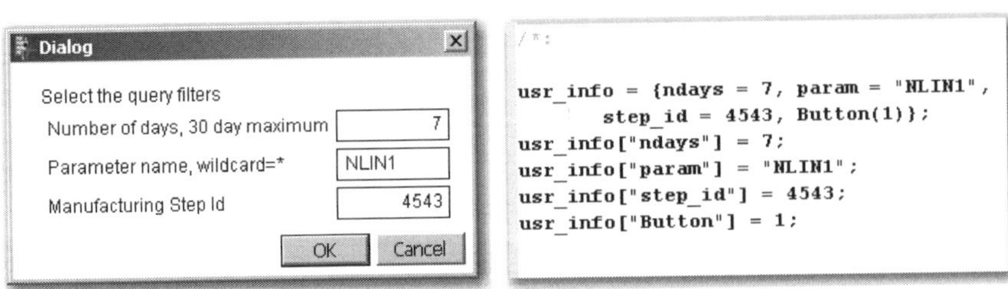

The user's responses are stored in a list named usr_info. In this list, all items but the last item, **Button(**1 | -1**)**, is an assignment statement. When the **Cancel** button is clicked, the value of Button is -1. When the **OK** button is clicked, the value is 1. Next add some statements to control program execution. If the user clicks **OK**, then proceed with the next set of steps. If the user clicks **Cancel**, then stop. Like JMP platform dialog boxes, most often clicking **Cancel** stops the entire analyses. We recommend the following commands or something similar for custom dialog boxes. These commands communicate to the user, clean up the caption box, and only proceed if the user clicks **OK**.

```
If(usr_info["Button"]== -1,
   Caption("Aborting..."); Wait(3); Caption(Remove); Throw() );
RemoveFrom(usr_info, NItems(usr_info));
//Remove the last item "Button(1)"
Eval List(usr_info); //Assignments are run
Show( usr_info, ndays, param, step_id);
```

The syntax to reference an item in the **Dialog( )** output usr_info is usr_info["Button"] or usr_info["ndays"]. The variable name on the left side of each assignment statement acts like a key to get the associated value from the usr_info list. (See Figure 7.2 (R).)

If **Button** is removed from usr_info, the rest of the items in usr_info form an assignment list. **Eval List( )** executes each assignment statement in the list. Thereafter, each item can be referenced simply as globals: ndays, param, and step_id. If **Cancel** is clicked, **Throw( )** is invoked, and the last three lines of code are not executed. At this point, you might want to review assignment lists in Chapter 5, "Lists, Matrices, and Associative Arrays." **Button** is always the last item in the list returned by **Dialog( )**. **RemoveFrom( )** requires the position of the item being removed (i.e., the last item = **N Items(** usr_Info **)**).

In the *JMP Scripting Guide,* Table 6.4 provides a synopsis of constructors for **Dialog( )** and **Column Dialog( )**. Or, you can try the examples in the online Help by selecting **Help ► JSL Functions ► Display**.

Built-in dialog box controls include **Edit Number**, **Edit Text**, **Radio Buttons**, **CheckBox**, **Text**, **ComboBox**, **ListBox**, and **Button** ("OK" | "Cancel").

Layout controls include **HList**, **VList**, **Lineup(**n, item1, ...**)**. **ColList** is available for **Column Dialog** only. These controls are similar, but are not the same as window controls **HListBox**, **TextEditBox**, and **NumberEditBox**, which are found later in this chapter.

Figure 7.4 Database Query Filter Dialog Box: (L) Using a ComboBox (R) Using a ListBox

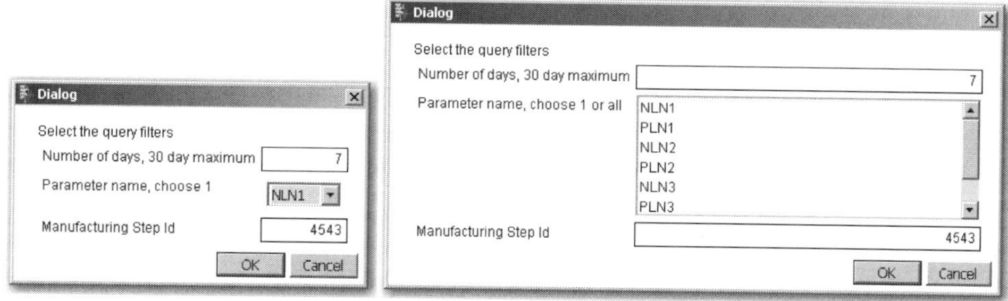

The 7_MessagesDialogs.jsl script includes examples and suggestions to try different options for dialog boxes. Figure 7.3 includes a dialog box using the **Edit Text** control for the user to type in the value for param. Figure 7.4 (L) shows the dialog box using a **ComboBox** that allows the user to select one item from a pull-down menu of parameter names. Figure 7.4 (R) uses a **ListBox** that allows the user to make multiple selections from the list of parameter names (through the SHIFT key and click or CTRL key and click method).

A Few Items to Note

- The 7_MessagesDialogs.jsl script provides examples of several controls for building a dialog box using the **Dialog( )** function.
- The dialog box created by the **Dialog( )** function is modal. It returns the user input in a list.
- The built-in dialog box has a limited set of controls. It might not be available in future versions of JMP.

Column Dialog Boxes

Data is the starting point for most analyses. So, it should not be surprising that the most frequently used dialog boxes include prompts to select data table columns, and then to cast those columns into roles for further analyses. The JSL utility function **Column Dialog( )** reduces this task to a simple function call.

The following script opens **Big Class.jmp** from the Sample Data directory, and generates a simple, familiar, user-friendly dialog box that has the same look as a JMP platform dialog box.

```
bc_dt = Open( "$SAMPLE_DATA/Big Class.JMP" );

col_dlg = Column Dialog(
    _yVar = ColList( "Y, Response" ),
    _xVar = ColList( "X, Treatment" ),
    _grpVar = ColList( "Group Factors" ),
    _wtVar = ColList( "Weight" ));
```

Figure 7.5 displays the results. On the left side of the dialog box, there is a list of columns from the current data table. For each **ColList** defined by your script, JMP generates a button and a list box to display user selections. Like a JMP platform dialog box, each **ColList** defines a role for upcoming analyses. Selecting a column on the left and then clicking a button casts the selected column to the specific role designated by the button. A user can also drag a column from the left and drop it into one of the list boxes.

The **OK**, **Cancel**, and **Remove** buttons are added by JMP without having to explicitly specify them in the script. You do not have control of the placement of these three buttons. The buttons and list boxes are aligned.

Figure 7.5 Select Columns Dialog Box

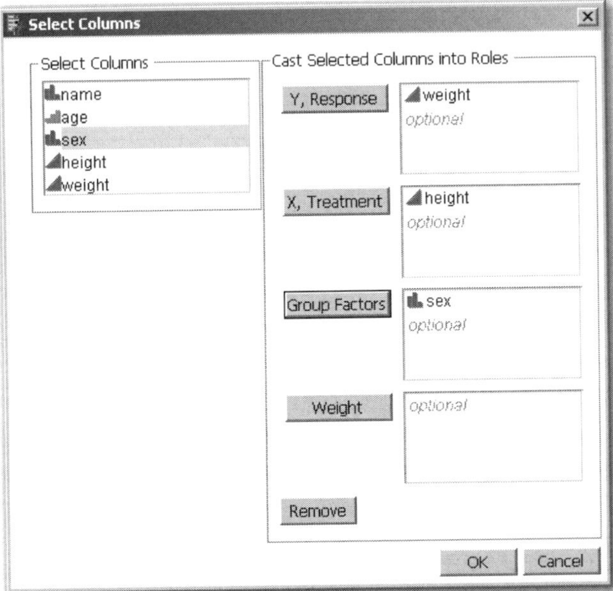

The dialog box created by **Column Dialog( )** is modal, which means it requires a user action before it proceeds.

If the columns are cast, and you click OK, the Log window shows the following:

```
col_dlg = {_yVar = {:weight}, _xVar = {:height}, _grpVar = {:sex},
  _wtVar = {}, Button(1)};
```

Column Dialog( ) returns a list of values in the same format as **Dialog( )**: a list of assignment expressions and Button (-1 | 1) as the last item. A result can be referenced by the assigned variable, col_dlg, and the quoted name for each expression. For example, col_dlg["_yVar"]) refers to the list {:weight} and col_dlg["_xVar"][1] refers to the column :height and col_dlg["Button"] is 1 (OK). If the last item Button, which is not an assignment expression, is removed, and an **Eval List** is applied, then the entire list of assignment statements is executed. Values can be referenced by the L-value of each assignment statement.

```
If (col_dlg["Button"] == -1, Throw( "User Cancel") );
Remove From( col_dlg , N Items(col_dlg) ); //Remove the last item
Eval List ( col_dlg );

Show( col_dlg, _yVar, _xVar, _grpVar, _wtVar );
```

Here is the general syntax for ColList:

```
ColList( "role", <MaxCol( n ) | MinCol( n )>, <Datatype( type )> )
```

Data type can be **Numeric**, **Character**, or **Rowstate**. When **MinCol (n)** is specified the user will be warned when pressing OK until the minimum number of columns has been specified (or the user cancels). **MaxCol(n)** limits the number of columns that can be specified for the associated role. The first argument for **ColList** can be any text string. However, long text strings are not very user-friendly. The 7_ColumnDialogs.jsl script includes a bizarre example for demonstration only that has a 1000-character string button. JMP can handle it, even though users will not!

In the *JMP Scripting Guide* , Table 6.4 provides a synopsis of constructors for **Dialog( )** and **Column Dialog( )**. **ColList** is available only for **Column Dialog( )**.

```
col_dlg = Column Dialog(
  _yVar = ColList( "Y, Response", MaxCol( 1 ), MinCol( 1 ),
      DataType( Numeric ) ),
  _xVar = ColList( "X, Treatment", MaxCol( 5 ), MinCol( 1 ),
      DataType( Numeric )  ),
  _grpVar = ColList( "Group Factors" ),
  _wtVar = ColList( "Weight", MaxCol(1) ),
   HList( "Confidence Interval Alpha", _alpha = EditNumber( .05 ) ));
```

Figure 7.6 (L) is the dialog box generated by this script. Additional controls in **Column Dialog( )** are always placed below the select columns area of the dialog box.

Figure 7.6 Select Columns Dialog Box: (L) From Code (R) With Custom Controls

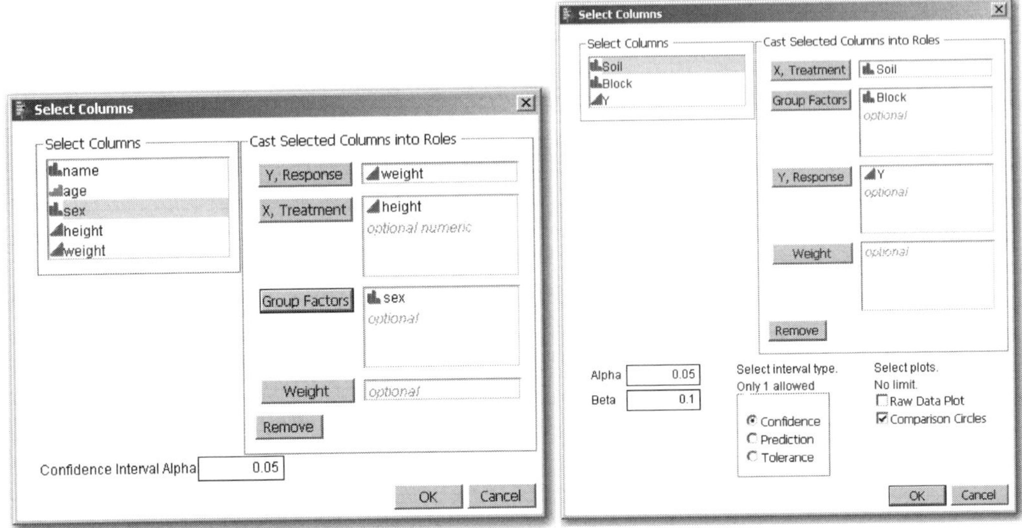

The 7_ColumnDialogs.jsl script includes the commands to create the custom controls in Figure 7.6 (R). It uses **VList**, **HList**, **Lineup**, and spaces for layout. It uses **Edit Number**, **Radio Buttons**, and **CheckBox** for additional controls. The placement of **Y, Response** does not follow the JMP standard and users might be confused. The JMP standard is to place **Y, Response** first. Avoid confusion by following the JMP standards when laying out controls.

A Few Items to Note

- If columns are **Hidden** in the data table, they are still available in **Dialog( )** and **Column Dialog( )** for casting. If columns are **Excluded** in the data table, they are not available for casting. Columns that are both **Hidden** and **Excluded** are not available.

- Strong names are always encouraged, but they are even more important for **Dialog( )** and **Column Dialog( )**. Look carefully at the L-values in the dialog box output. There is no double-colon prefix-scoping for the variables.

  In the first example script in this section, suppose the name _wtVar was replaced with weight, the name of the role weight. If you run the JSL statements to remove Button and apply **Eval List**, JMP throws an error to the Log window if you are using the Big Class data table or any table that has a column named weight. (The error is shown at the end of this section.) This is a name collision: both a variable and a column have the same name.

Remember, if JMP cannot resolve a name, it looks for a global variable. (The global variable weight does not exist until **Eval List** is run.) If JMP cannot find a global variable, then it looks for a column with that name. JMP finds a column, but it does not recognize **Eval List** for a column. Using leading underscores and abbreviations builds stronger names because it reduces the likelihood of a name collision. See the section "Using Namespaces" in Chapter 9, "Writing Flexible Code," for more discussion about name collisions.

```
Cannot set value for the column 'weight' because the row
number (-1) is not valid.
```

Dialog Boxes Using the New Window Function

In addition to the display box objects and the functions introduced in Chapter 6 to organize and customize reports, **New Window( )** supports display box objects and functions to communicate with users. Some of these share the same names as **Dialog( )** function controls (for example, **CheckBox** and **ComboBox**). Others have the word "box" added to their names (for example, **ColListBox** and **LineupBox**). The look and feel are familiar to the user. Yet, these display box objects for communication have different syntax and extended features.

Dialog boxes for communicating with users have three main tasks:

1. Define the interface (the layout of controls and the look and feel).
2. Get (also known as unload) user input.
3. Check information and use it to control the logic of the program.

The simplest example of controlling program logic is halting the script if **Cancel** is selected. Remember these three main tasks while you explore display boxes to communicate with users.

The first block of statements in the 7_DialogsNewWindow.jsl script creates and compares two dialog boxes. The dialog box dlg_usr is created with Dialog, and win_usr is created with **New Window**.

Run the 7_DialogsNewWindow.jsl script, look at the Log window, and select the topic Help. The syntax of **CheckBox** is a simple assignment statement in which the user is allowed to change the setting by clicking on the check box once it is displayed. When the user clicks OK, a list of control settings is returned.

The **CheckBox** display box object used in the script with **New Window( )** has several features. **Enable(0 | 1)** is one of these features, and allows the script to gray out this control. The general syntax is **CheckBox({list}**, opt1, opt2, <*script*>**)**. The check box for **New Window( )** is slightly more complex than **Dialog( )**, but its features simplify complex modal and non-modal scenarios. The script in a **CheckBox** display box object is run when a user interacts with the display box. (This allows for an immediate response before the user clicks **OK**). Here is a simple, two **CheckBox** example:

```
//-----------Create a simple UI dlg_usr with Dialog function--------
sp = "   ";
dlg_user = Dialog( VList( "Select options and click OK",
  V List( sp, "Normal Quantile Plot",
   Lineup(2,
     cb1 = CheckBox("Plot Actual by Quantile",0), sp,
     cb2 = CheckBox("Plot Quantile by Actual",1), sp)
   ), sp,
   Hlist (Button("OK"), Button("Cancel"))
));

//-----------Create a simple UI win_usr with New Window function-----
win_usr = New Window("Select Options and click OK", <<Modal,
  VListBox(
    TextBox("Select Options and click OK"),
    SpacerBox(Size(0,15)),
    TextBox("Normal Quantile Plot"),
      SpacerBox(Size(0,5)),
         cb1 = CheckBox({"Plot Actual by Quantile"}, <<set(1,0) ),
         cb2 = CheckBox({"Plot Quantile by Actual"}, <<set(1,1) )
       ), SpacerBox(Size(0,20)),
       HListBox(SpacerBox(Size(60,1)),
       ButtonBox("OK",
       ), //end HListBox
       Button Box("Cancel"))
);
```

Figure 7.7 Check Box Dialog Using: (L) Dialog Function (R) New Window Function

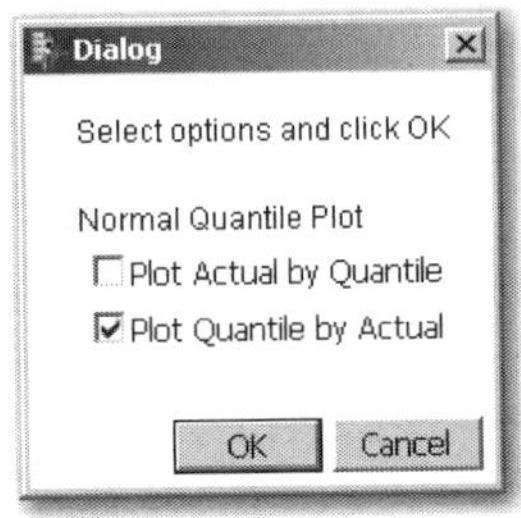

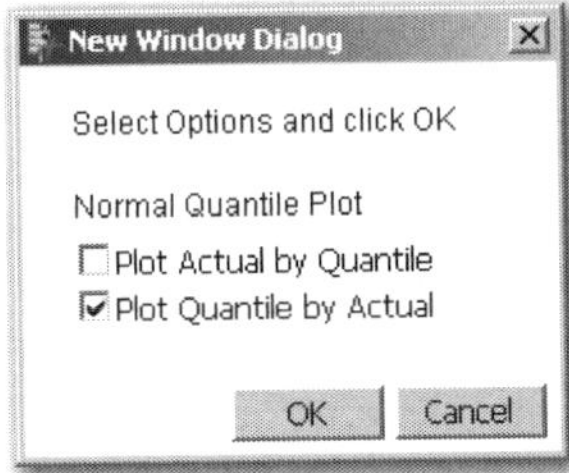

Run each block of the previous code one at a time, and then look at the Log window. Let's catalog the differences of the second block of code in the script. **<<Modal** is declared to emulate the **Dialog( )** function's behavior. Instead of lining up spaces, **SpacerBox( )** is used to add horizontal and vertical spacing. Each **CheckBox( )** has two arguments: a list with one item, and the command to pre-load either checked (value of 1) or unchecked (value of 0).

Dialog( ) returns the list of settings when the user clicks **OK** or **Cancel**. A modal **New Window( )** returns only the value of the button (1 for **OK** and -1 for **Cancel**). Note this rule: when creating a **New Window( )** dialog boxes, your script needs to retrieve the user information.

For this simple modal dialog box example, a script can be added to the OK button to retrieve the values in the display box objects. The next section, "Retrieve User Input," includes **Get( )** syntax, options, and recommendations. The following is a snippet from a block of commands that has a script attached to each **CheckBox( )** and **ButtonBox( )**. After you run the block of commands, a display box-specific message is displayed in a caption window. The message changes when the user interacts with the display box controls.

```
// Dialog with attached scripts at each checkbox and button
win_test = New Window("Test Attached Script", <<Modal,
        VListBox(
        TextBox("Select Options on/off then click OK"),
        SpacerBox(Size(0,15)),
         a_cb = CheckBox({"Check A"}, <<set(1,aVal),
              aVal = setit(aVal, "A")  ),
         b_cb = CheckBox({"No Check Me, B"}, <<set(1,bVal),
          bVal =setit(bVal, "B")
         ) . . .
```

A display box script can do much more than manage a caption. Typically, a display box script includes commands to unload settings, change the display, or direct the program to do something.

The 7_DialogsNewWindow.jsl script contains code to create the familiar-looking dialog boxes in Figure 7.8 (dlg2_usr and win2_usr, respectively). The script for win2_usr is an explicit, brute-force version. Functions and associative arrays can reduce the repetitive JSL statements.

To summarize, using **Dialog( )**, the scripter defines the layout with a limited set of controls. This function returns a list of all settings and closes the dialog box after a button is clicked. Using **New Window( )**, the scripter defines the layout and actions with an extended set of controls. It is the scripter's responsibility to unload each control's settings and close the dialog window. The additional coding burden is offset with the opportunity to create dynamic complex displays that direct program logic.

Figure 7.8 Check Box Dialog Boxes Using: (L) Dialog Function (R) New Window Function

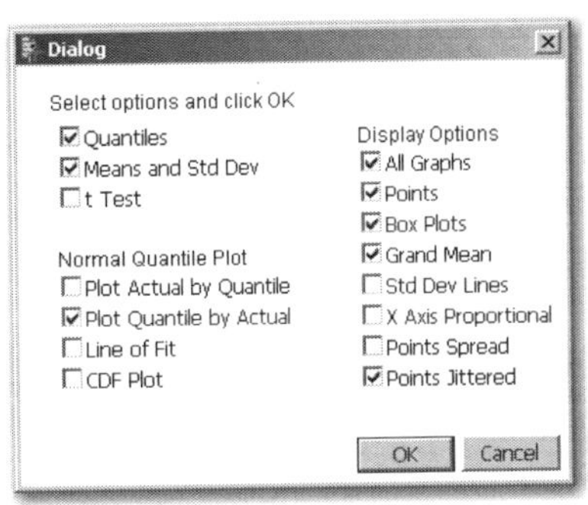

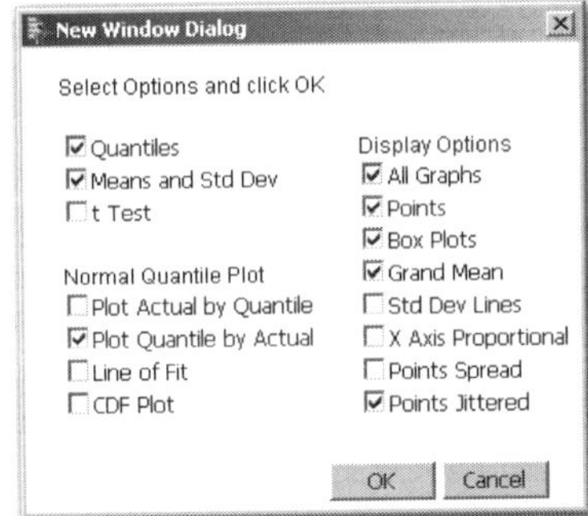

Retrieve User Input

To make a **New Window()** dialog box useful, you need to know how to retrieve the user-entered information. There are many interactive display box objects that can be used in a **New Window** dialog box, but only a few messages are needed to retrieve the user's response. The key messages are **Get**, **Get Text**, **Get Items**, and **Get Selected**. The following table describes these messages and a few others that are commonly used when working with interactive display boxes.

Table 7.1 Messages to Retrieve User Input

Message	Use	General Syntax
Get	Retrieves values for interactive display box objects	var = display_box << Get
Get Text	Retrieves the text from a TextBox or a TextEditBox.	var = display_box << Get Text
Set Selected or **Set**	Allows the preselection of an item such as a ColList or ListBox. Sets the initial value of an item such as CheckBox or RadioBox.	display_box << Set Selected (item number, state)
Get Selected	Returns the value of a selected item.	var = display_box << Get Selected
Remove Selected	Removes the selected items from the current list in the display box.	var = display_box << Remove Selected
Get Selected Indices	Returns the index number of a selected item.	var = display_box << Get Selected Indices
Get Items	Retrieves a list of selected items from a ColListBox or ListBox.	var = display_box << Get Items

The syntax is very similar for all of the messages. The key to retrieving user information is to create a reference to each display box of interest. In most instances, each interactive display box should have a name, a variable reference to the control, and a variable reference to its selected values.

You should initialize all expected responses. As discussed in the 7_IntroDialogs.jsl script, just because the user has a prompt window with controls does not guarantee a user response. Also, the retrieval messages listed in the previous table are typically included in the script attached to a display box control. The message is run only when the user interacts with the display box. Without an initial value, it is easy to create a scenario where the **Get** message is not run, and the subsequent analysis does not capture and handle empty results.

Example Using Get, Get Text, and Set

The script for this section is 7_RetrievingInput.jsl. Consider the first example, where information is contained in a **TextEditBox**, a **NumberEditBox**, and a **RadioBox**. This information needs to be retrieved. This example retrieves all of the information when the user clicks **OK**.

```
//Modal Retrieval in OK button
//Initialize
genderList = {"Male", "Female"};
user_name = "Last, First"; neb_a = 29; rb_g = 2;

New Window( "Retrieve Info", << Modal,
  Text Box( "Please provide your name, age and gender" ),
  Text Box( " " ),
  HListBox(
    Panel Box( "Personal Information",
      Lineup Box(2,
        Text Box( "Name:" ),
        teb_name = Text Edit Box( user_name ),
        Text Box( "Age:" ),
        neb_age = Number Edit Box( neb_a ),
        Text Box( "Gender: " ),
        rb_gender = Radio Box( genderList , <<Set( rb_g ) )
      ) // end LineupBox
    ), //end PanelBox
    Panel Box( "Action",
      VListBox( ok_btn = Button Box("OK",
         user_name = teb_name << Get Text;
         neb_a = neb_age << Get;
         rb_g = rb_gender << Get;
         show(user_name, neb_a, rb_g, genderList[rb_g],
           // Get Selected retuns the list value
          (rb_gender << Get selected) )
      ), //end OK
         cncl_btn = Button Box ("Cancel")
      )
) ) );
```

A Few Items to Note

- If you remove the expression <<Modal from this script, its behavior looks like a modal dialog box waiting for a response. Yet, before clicking **OK** or **Cancel**, try opening a file in JMP, or any other action for that matter. It is not modal! Enter a new value for age, do not click **OK** or **Cancel**, and in the Log window, type the command **show(** neb_a). The value has not been updated. Now, click **OK**. The Log window reveals the new value. By placing all **Get** statements (or any subsequent action) in the **OK ButtonBox** code, you create what we call a *pseudo-modal* dialog box.
- This script follows our recommendations. Result variables are initialized, and each control is named and has an associated result variable. teb_name refers to the TextEditBox collecting name information, and the result is user_name. rb_gender refers to the RadioBox for gender, and the user-selected index is rb_g. neb_age refers to the NumberEditBox for age, and neb_a is the associated result value.
- For non-modal dialog boxes, commands to close the dialog window should be attached to the **OK** and **Cancel** buttons. For simplicity, the previous script did not include these commands. Other scripts provide examples.

Most, but not all interactive display boxes allow a script to be attached. The attached script runs when a user interacts with the display box. For the **TextEditBox**, the user types a response, and when he or she presses the ENTER key on the keyboard, the script runs. Unfortunately, a script cannot be attached to a **NumberEditBox** in JMP 9.

There are a couple of alternate solutions. The following script uses a **TextEditBox**, and when the user enters a value, the script checks whether a numeric value was entered. If it was not, the **TextEditBox** is emptied, which is the expected response when a user enters an invalid value in a field. If the user's response resolves to a number, then **neb_a** is set to the numeric equivalent.

This script requires a few extra lines of JSL, but it enables a completely interactive display box without an extra button to update values. Using a **GlobalBox** is another alternative, and it is presented later in this chapter.

```
//Non-Modal Retrieval in Display Box Scripts no OK/Cancel
//Initialize
genderList = {"Male", "Female"};
neb_name = "Last, First"; neb_a = 29; rb_g = 2;

New Window( "Retrieve Info",
  Text Box( "Please provide your name, age and gender" ),
  Text Box( " " ),
  Panel Box( "Personal Information",
Lineup Box( 2,
  Text Box( "Name:" ),
```

```
   teb_name = Text Edit Box( "Last, First",
     <<Script( user_name = teb_name << Get Text )
   ),   //end teb_name
   Text Box( "Age:" ),
   teb_age = Text Edit Box( char(neb_a),
      << Script(
         If( !Is Missing(Num(teb_age << Get Text )),
           neb_a = Round(num(teb_age << Get Text), 0);
           teb_age << Set Text( char(neb_a) ),
           teb_age << Set Text ("") );
           Show( neb_a)
    ) // end teb_age script
   ),  //end teb_age
     Text Box( "Gender: " ),
rb_gender = Radio Box(genderList , <<Set( rb_g ),
      <<Script(
         rb_g = rb_gender << Get Selected
        )
    )   // end rb_gender
   )  // end LineupBox
   ) //end PanelBox
);
```

Example Using Get Selected, Remove Selected, Get Items

This example shows how to create a **ColListBox**, get the selected columns, and add them to a list. In addition, it removes selected items from the list. It uses the Sample Data table Fitness.jmp. This script is a modification to the example in the *JMP Scripting Guide* for ColListBox. It uses an **IfBox** to turn off the display if no columns are selected, and it restructures the code to be more modular using expressions for the attached scripts.

```
//SECOND SCRIPT:  This shows how to select and remove columns
//
clear globals();
orig_dt = Open( "$SAMPLE_DATA/Fitness.jmp" );

cnt = 0; Chosen Columns={};  //initiate values

//----Action to take for Add------------------------------
add_script = Expr(
   listocols << Append( selcol << Get Selected );
   // Send Get Items to a Col List Box
   Chosen Columns = listocols << Get Items;
   cnt = N Items(Chosen Columns);
   showIf << set (cnt)
);
```

```
//----Action to take for Remove ----------------------------
remove_script= Expr(
   listocols << Remove Selected;
   // Send Get Items to a Col List Box
   Chosen Columns = listocols << Get Items;
   cnt = N Items(Chosen Columns);
   showIf << set (!!cnt)
);

//-----Dialog Box layout----------------------------------
dist_win = New Window( "Create Distributions",
  H List Box(
    selcol = Col List Box( All, width( 100 ), nlines( 6 ) ),
     Lineup Box( N Col( 1 ), Spacing( 3 ),
        Button Box( "Add Column >>", add_script ),
        Button Box( "<< Remove Column", remove_script )
     ),
  // listocols is a Col List Box
  listocols = Col List Box( width( 100 ), nlines( 6 ) )
  ),
 Text Box( " " ),
 // Show what Get Items returns
    showIf = IfBox(0, stuff = Global Box(Chosen Columns ))
);
```

Figure 7.9 Custom Dialog Box Using ColListBox and IfBox

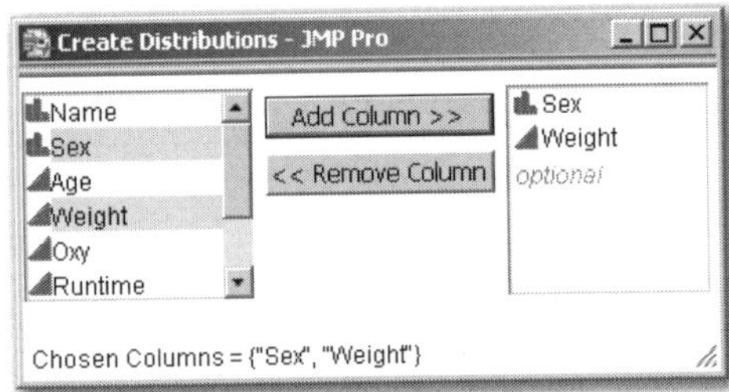

A Few Items to Note

- **GlobalBox( )** did not require a **Set** command to update the display. The section "Interactive Displays" contains additional information and an example of this often-used display box.
- After adding several columns, remove all columns. Chosen Columns disappears. showIf is an **IfBox( )**. It is used often. However, first attempts to use this display box object and other display box objects are not always successful. The key to success is to recognize the general pattern and practice with the examples for each display box. See the online Help by selecting **Help ▶ Display Box Scripting**. The general pattern is that display boxes are containers. They have initial values, and most of them need to have a message sent to update the contents. In other words, they do not act like **GlobalBox( )**. The values and contents are not the actual box. Hence, you should maintain both a display box reference and a variable to represent its contents for each communication display box used in a dialog box. When

creating a pseudo-modal dialog box, where no user information is retrieved until the user clicks **OK**, some scripters repurpose the variable name for the control to be the values (for example, cb = cb << Get(1)). This is valid if the script for the **OK** button also closes the dialog box. If you are new to scripting or you script infrequently, we recommend using a different variable name (for example cbset = cb << Get (1). It is easier to read and easier to convert your script from pseudo-modal to interactive, if or when there is a need to do so.

- Retrieving information from a **New Window( )** dialog box requires much more scripting than the functions **Dialog( )** and **Column Dialog( )**. However, **New Window( )** enables dynamic displays and drill-down methods. A dynamic interactive window requires display statements and retrieval statements, and can also require update statements. User friendly interactive displays usually require deep nesting of display boxes and actions. We recommend that interactive display box controls be defined using expressions and functions before the **New Window( )** layout. Your script will be more readable and extensible.

Interactive Displays

Previous sections in this chapter described the general pattern for creating interactive displays using **New Window( )**, layout display boxes, and interactive display boxes, The sections included very simple examples of getting, setting, and updating information. This section describes two special display boxes, a **Slider Box( )** and a **GlobalBox( )**, and provide additional examples of interactive displays.

The **SliderBox( )** and **GlobalBox( )** are used in a script included for Chapter 4, "Essentials: Variables, Formats, and Expressions." This script demonstrates the use of expressions. Figure 7.10 displays a graph of three Weibull probability density functions with parameters that can be changed by using either a **SliderBox( )** or by typing them in a **GlobalBox( )**. A **ButtonBox( )** allows the user to request Weibull random data based on the specified parameters. This type of interactive script is useful as a training tool.

Figure 7.10 GlobalBox, SliderBox, and ButtonBox Interactive Display

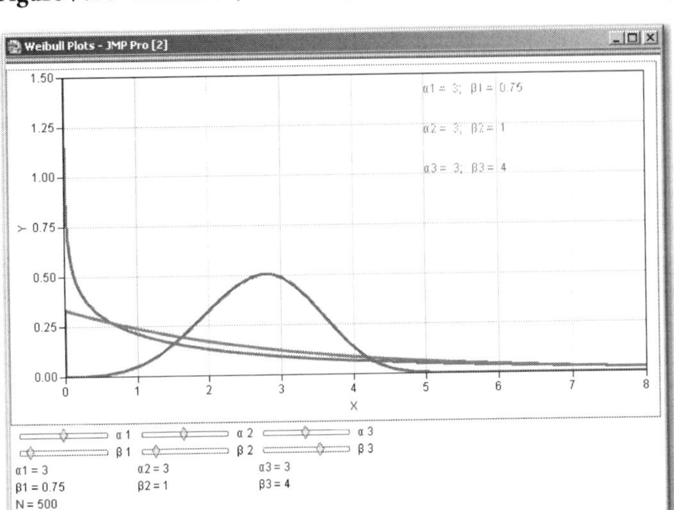

A **SliderBox( )** draws a slider control that the user can move to select the value within a range. Here is the general syntax of a **SliderBox( )**:

```
SliderBox(min, max, global, script)
```

A **GlobalBox( )** shows the value and allows the user to edit a current JSL global variable. Here is the general syntax of a **GlobalBox( )**:

```
GlobalBox(global)
```

A **ButtonBox( )** draws a button with the specified name that executes a script when the button is clicked. Here is the general syntax of a **ButtonBox( )**:

```
ButtonBox("Button Name", script)
```

The 7_InteractiveDisplays.jsl script includes a simple example of these three controls. This script creates an interactive display that prompts the user to select a score from 1 to 100. The selection can be made by using either a slider bar or by typing the value in a global box. Once the score is defined and the **Score** button is clicked, the score is displayed using a **Caption( )**. This demonstrates retrieving information from a user and using it later in a script. Note that sb and gb are display box controls that share a common global value, score, that does not require a message to be automatically updated.

```
score = 50;
New Window( "Judge the Project",
  Panel Box( "Select score on a scale from 1 to 100",
    sb = Slider Box( 1, 100, score ),
    sb << Set Width( 250 ),
    gb = Global Box( score ),
    Button Box( "Score",
      Caption( "The final score is " || Char( score ) );
      Wait( 3 );
      Caption( Remove ))
));
```

Previous sample scripts from this chapter included examples using **CheckBox**, **RadioBox**, **NumberEditBox**, **TextEditBox**, **ColListBox**, **ButtonBox**, and **IfBox**. There is more information if you need it. The syntax and demo scripts for these display box objects are found in the online Help by selecting **Help ► Display Box Scripting**. The remainder of this section provides additional examples. The first script extends the previous section's script to select and remove items using **ColListBox** to create Distribution plots of the selected columns.

Figure 7.11 Custom Dialog Box Using ColListBox and Embedded Distributions

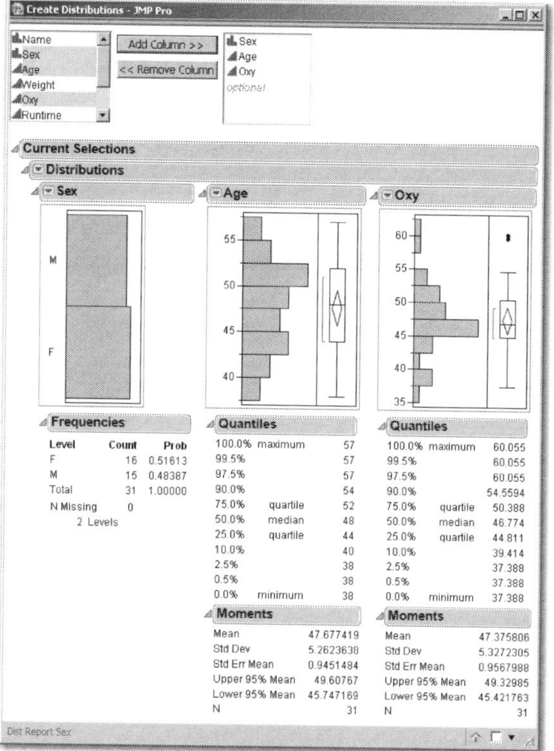

Only two changes were needed to achieve this interactive dialog box. The first change was **IfBox( )** was replaced with result_ob = Outline Box("Current Selections"). The second change was to the refresh_script expression. It was replaced with the following:

```
//----Refresh display script---------------------------
refresh_script = Expr(
   Try(dist_vlist << delete);
   If( (cnt > 0 & N Items( Chosen Columns ) >= 1),
   show(Chosen Columns);
   dist_vlist = V List Box(
   orig_dist = orig_dt << Distribution( Column( Eval(Chosen Columns)
       ), invisible )
    );
   result_ob << append( dist_vlist );
));
```

Another alternative script can be created by modifying the refresh_script expression. The alternative script could create an interactive display that retains the previous selections with their outline boxes closed.

The 7_InteractiveDisplays_ChoiceExperiment.jsl script creates a dialog box that requests input from the user on preferences for a personal laptop. This is the same set of questions provided in the *JMP Design of Experiments Guide* in Chapter 8, "Discrete Choice Designs." The script asks for answers to questions using **TextEditBox**, **NumberEditBox**, and a **RadioBox**. The user is asked to provide preferences for eight choices related to four attributes of a personal laptop. Once the user has answered the questions, a **ButtonBox** allows the user to send the results to a JMP table.

The 7_Extra_InteractivePopupTrendrulesTest.jsl script is an example of a dialog box with a **PopUpBox** control. It was created for 7_Extra_ControlChartInterface.jsl, which is a prototype interface for a custom application. Both scripts are included as concept scripts for creating a user-friendly user interface.

Design a Dialog Box

Dialog boxes and interactive displays become more important when a script is distributed to many people, when the analysis calls for the user to direct the analysis path, or when building a drill-down capability.

When it comes to building an application, it is necessary to scope the problem. Know your users' skills, their environment, and their usage. Know the number of variables, number of rows, frequency, data sources, data issues, etc. This step is called gathering requirements or creating a specification.

After scoping the problem, if a dialog box is needed, and you are trying to determine the layout and which controls to use, our many recommendations boil down to a few. Look for good templates; design the dialog box; collect user feedback; prototype and test each interface; collect user feedback; modify, integrate, and test; collect user feedback; monitor usage; and periodically collect user feedback or provide a mechanism to gather issues and suggestions.

When it comes to coding the application, know your environment, know your tools, and follow our recommended guidelines for working with JMP windows and interactive displays. Also, break down tasks into smaller steps.

Often, watching someone perform a task is the fastest way to learn. The remainder of this section builds a dialog box that shares features of the **Fit Y by X** dialog box.

Figure 7.12 Fit Y by X Dialog Box

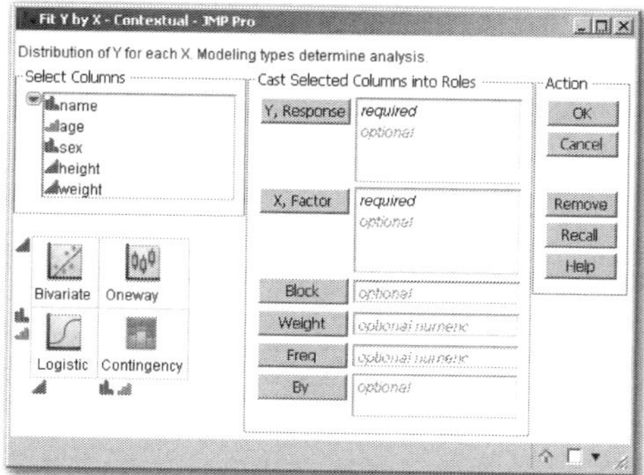

Dialog Box Building Exercise—Know Your Tools

For this exercise, ignore the graphics in the bottom left area of the **Fit Y by X** dialog box in Figure 7.12. To build the dialog box in Figure 7.12, the display box objects used include 1 TextBox, 3 PanelBox, 11 ButtonBox, 7 ColList Box, 1 PopUpBox, 2 or more LineupBox, 1 HListBox, 2 VListBox, and about 3 SpacerBox. In addition, there are scripts that define the behavior when a user interacts with any of these controls.

Step 1: Create the Layout—Framing

Study and run the 7_DesigningADialogWindow_Step1.jsl script. Look up any display box object that is unfamiliar.

```
_win_info = "Distribution of Y for each X. Modeling types "
      ||" determine analysis." ;

//-- step 1  create the layout - framing
New Window("Fit Y by X - Contextual - Step 1",
```

```
VListBox(
  TextBox(_win_info),
  SpacerBox(Size(0,5)),
  HListBox(
    PanelBox("Select Columns", ),
    PanelBox("Cast Columns into Roles",),
    PanelBox("Action")
  ) // end HListBox
 )  //end VListBox
);  //end New Window
```

Figure 7.13 Step 1: Laying Out the Fit Y by X Dialog Box

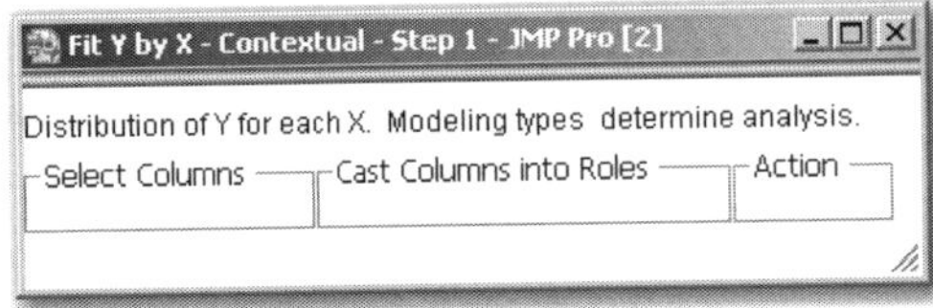

Step 2: Create Display Box Controls and Initialize Results

Keep in mind that good, strong names and proper scoping are the first line of defense against name collisions. See the section "Using Namespaces" in Chapter 9. The following commands should precede the **New Window( )** statement. As you test, ensure that at least one data table is open. This script contains explicit, brute-force coding for **ColList Box( )** and **Button Box( )** because this is a training script. The display built by step 2 (see Figure 7.14) should look familiar.

```
//initialize
_yvarList={}; _xvarList={}; _blkList={}; _wtList={};
_freqList={}; _byList={};

//--Cast into Roles Controls
_inputCol = ColListBox(ALL, width(100), nlines(9)); //all current
data table columns

_yvar_clb = ColList Box(width(100), nlines(6), min items(1) );
_xvar_clb = ColList Box(width(100), nlines(6), min items(1) );
_blk_clb  = ColList Box(width(100), nlines(1) );
_wt_clb = ColList Box(width(100), nlines(1), max items(1), numeric );
_freq_clb = ColListBox(width(100), nlines(1),max items(1), Numeric);
_by_clb = ColListBox(width(100), nlines(2));

_yvar_bb = Button Box( "Y, Response",  );
_xvar_bb = Button Box( "X, Factor",  );
_blk_bb = Button Box( "Block",  );
_wt_bb = Button Box( "Weight",  );
_freq_bb = Button Box( "Frequency",  );
_by_bb = Button Box( "By",  );

//--Action Controls
_ok_bb = Button Box("OK",  );
_cncl_bb = Button Box("Cancel",  );
_rmv_bb = Button Box("Remove", );
```

```
_redo_bb = Button Box("Recall", Enable(0) );
_redo_bb << Set Tip("Not implemented yet");
_Help_bb = Button Box("Help",  );
```

Test the built-in **Fit Y by X** dialog box. **Weight** must be numeric, and only one column can be used to define **Weight**. The remaining **ColListBox( )** attributes are based on the JMP interface.

Add a **LineUpBox( )** to align each **ButtonBox** and **ColListBox** in the **Cast Columns into Roles PanelBox**.

```
Panel Box("Cast Columns into Roles",
 LineUpBox(2,
   _yvar_bb, _yvar_clb,
   _xvar_bb, _xvar_clb,
   _blk_bb,  _blk_clb,
   _wt_bb,   _wt_clb,
   _freq_bb, _freq_clb,
   _by_bb,   _by_clb
 )
),
```

Add a **VListBox** of button boxes to the **Action PanelBox**.

```
PanelBox("Action",
  VListBox(_ok_bb,
    _cncl_bb,
    SpacerBox(Size(0,20)),
    _rmv_bb,
    _redo_bb,
    _Help_bb
  )
```

Figure 7.14 Step 2—Create Display Box Controls Using Fitness.jmp

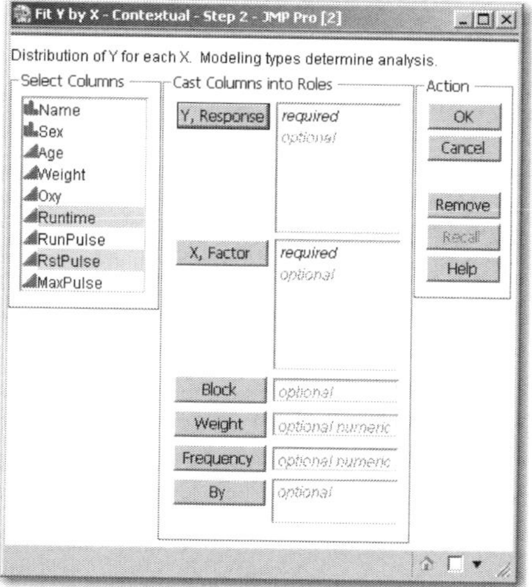

Run the 7_DesigningADialogWindow_Step2.jsl script. You can select current data table columns, but they cannot be cast into roles until the five role buttons have an attached script. A single function or expression works for casting roles.

Review the script commands. Can you spot some obvious program style improvements? The most obvious is to create a variable for the **ColListBox** width. If a global variable defined list box widths, and user reviews revealed that wider list box displays were needed, only a single command would need to change, instead of finding all instances of width and changing them.

Step 3: Create Functions and Expressions and Add to Buttons

The 7_DesigningADialogWindow_Step3.jsl script contains the full script. Only key snippets are presented here. The single function for casting a role is _addCol_fn(target_clb). The JMP **ListBox( )** messages, **Append** and **Get Selected**, make this a simple task. These messages were introduced in the section "Retrieve User Input." The **OK** button script needs to do the most work. However, by breaking down the work into smaller steps, four simple commands define what happens. Because **OK** and **CANCEL** both require closing the dialog box, write the command once, and assign it a simple name.

Another level of abstraction is relatively easy to accomplish. For example, use lists or associative arrays for display box references and options (Min(n), numeric, newlines, etc.) and display box settings. Then, a **For** loop could replace six varList initializations, six varList assignments, six role **ButtonBox** definitions, etc.

If you are new to scripting or you script infrequently, often explicit, brute-force scripting is easier to read today and two months from now. In the following script, **----)** represents missing JSL statements.

```
//Since a single source_inputCol, only the target_clb is needed
_addCol_fn = Function( {target_clb}, {Default Local},
    target_clb << Append( ::_inputCol << Get Selected );
// Send Get Items to a Col List Box
);

//-- close window
_closeit = Expr ( Current Window() << close window() );

//-- return button status
_btn_fn = Function({val},
      If( val!=1, val = -1);
      Button = val   );

//-- cancel
_cancel_xp = Expr( _btn_fn(-1); _closeit );
```

```
//-- get results
_get_rslt_xp  = Expr(
    _yvarList = (_yvar_clb << get Items);
    _xvarList = (_xvar_clb << get Items);
----);

_ok_action_xp =Expr(
    _get_rslt_xp;  //get results
    _btn_fn(1);    //set the button response
    _closeit;      // close this window
Show( _yvarList, _xvarList, _blkList, _wtList, _FreqList,
   _byList, Button)
);

//-- remove items
_remove_xp =Expr(
    _yvar_clb << Remove Selected;
    _xvar_clb << Remove Selected;
----);
```

Now, add these functions or expressions to the buttons. Here are just two examples:

```
_yvar_bb = Button Box( "Y, Response", _addCol_fn( _yvar_clb ) );
_xvar_bb = Button Box( "X, Factor", _addCol_fn( _xvar_clb ) );

//--Action Controls
_ok_bb   = Button Box("OK", _ok_action_xp );
_cncl_bb = Button Box("Cancel", _cancel_xp );
```

A Few Items to Note

- Command of JSL to manipulate lists, expressions, and functions (in particular, lists) is the bridge to next-tier scripting.
- Expressions and functions do not run until they are executed. When a button is clicked, the attached script (function or expression) is run.
- The **Web( )** function is used to call JMP online Help. It is a utility function only on Windows. See Chapter 9 for more details.
- **Current Window( )**, like **Current Data Table( )**, is a useful command.
- A statement was added at the beginning of the previous script to test whether the number of open tables was zero. If it was zero, the user was prompted to select a file. If the user did not open a file, the script stopped. The function **N Table( )** returns the number of open tables in the current JMP session.
- Output from the custom dialog box includes six lists and buttons. After running this script, making your selections, and clicking **OK**, see the Log window for results.
- Building the JMP **Fit Y by X** dialog box is as easy as 1, 2, 3. As a bonus, two additional items were added to this dialog box in the 7_DesigningADialogWindow_Extra.jsl script. An

approximate replica of the **Fit Y by X** graphic was added using **IconBox( )**. And, a **PopupBox( )** was added that looks like the JMP Column Filter. This last item was added as a concept script, demonstrating the potential use of **PopupBox( )**. It is not fully functional.

Deploy User Input

A subtitle for this section could be "Summary of Communicating with Users and Segue to Building Custom Displays."

Put It All Together

This chapter introduced JMP built-in dialog boxes, utility message functions, and display box objects to gather and send information to the user. Several scripts demonstrated modal and non-modal communication dialog boxes and how to retrieve information.

Several of the sample scripts included examples of deploying user input to draw Weibull plots, create a table of user survey responses, embed active JMP Distribution reports upon request, and direct the remaining program flow based on user response.

This section discusses the steps to integrate the information gathered from the dialog box with analyses.

Code Structure

If your application requires user input, the layout of your code depends on whether your dialog boxes are modal, pseudo-modal, or interactive. Assume that only one initial dialog box is required. Pseudocode (high-level description of code) is provided in this section to describe the different options.

A Few Items to Note

- For built-in dialog boxes, all of the analysis action occurs after the user clicks **OK**. The analysis code (or at least the statement that calls up the analysis expression, function, or file) must follow the dialog box. For modal **New Window( )** dialog boxes, the code or analysis action can occur after the New Window command, or can be called as an action for the OK button.
- For pseudo-modal or interactive dialog boxes, the **New Window( )** command that defines the dialog box should be the last statement in the file. Statements that follow are likely to complete before the user interacts with the dialog box. Some scripters like to take advantage of this fact, and they assume that code will run before the user even sees it. So, scripters

define the analysis actions after the **New Window( )** command. This approach has caused confusion that can result in a broken script.

- It is recommended for modal **New Window( )** dialog boxes (especially the ones that have analyses code following the dialog box) that you create a global container, list, or an associative array or global expression that contains the results (user settings). It does not have to be an assignment list like those returned by built-in dialog boxes. This allows a single statement to see the user settings, and it helps in testing and debugging. It might be useful for an experienced scripter to use the Window namespace. (See Chapter 9.)
- Fully interactive dialog boxes require actions to be taken when the user interacts with an edit box, a slider, a global box, a button, or even a change in a list box entry or a new option from a pop-up box. Make sure that the control has the ability to attach a script or to build a button to signal the program to get the new information and do something!

If you are writing **New Window( )** dialog boxes (especially non-modal), the most reliable coding method is to define all actions as expressions or functions before the dialog box.

Modal Built-In Dialog Box

```
<Modal Dialog() | Column Dialog() only restricted list of controls
     Dialog:
       usr = Column Dialog( ColList, Other Controls, OK);
       If ( Usr["Button"] == -1, Throw() );
     //else

     Analysis Code:
        A set of instructions | Include () | Expr | Function(usr)
     //end program
/Modal Dialog>
```

Modal New Window Dialog Box

```
<Modal New Window("title", << Modal, OK )

     Define Dialog controls:
        for each control create 2 references object, settings
        cb_set = 0;  cb_obj = CheckBox(...)
        _yvarList = {};  _yvar_clb = ColListBox();

     Define Actions:
         closeit = Expr( current window() << Close window() );
            task_k = Expr| Function(code etc )

     Define Analysis <optional> : Expr or series of Expr | Function

     Dialog:
       usr = New Window( "title", << Modal,... OK);
       if ( usr == -1 , Throw() );
     //else

     Analysis Code: <required if OK action did not include analysis>
```

```
      A block of instructions | Include () | Expr | Function(usr)
/Modal New Window>
```

Pseudo-Modal New Window Dialog Box

```
<Pseudo-Modal New Window("title", OK): user has access to JMP UI
     Define Dialog controls:
       for each control create 2 references object, settings
       cb_set = 0;  cb_obj = CheckBox(...)
       _yvarList = {};  _yvar_clb = ColListBox();

    Define Actions:
        closeit = Expr( current window() << Close window() );
        task_k = Expr| Function(code etc )

    Define Analysis | Include: Expr or series of Expr | Function

    Dialog:
       New Window ("title", ...,
          Cancel: Throw(); Closeit;
          OK: Analysis | A block of instructions | send results
          Closeit
        );  //end new window
/Pseudo-Modal>
```

Interactive New Window Dialog Box

```
<Interactive New Window("title", ...): user has access to JMP UI
     Define Dialog controls:
       for each control create 2 references object, settings
       cb_set = 0;  cb_obj = CheckBox(...)
       _yvarList = {};  _yvar_clb = ColListBox();

    Define Actions:
        closeit = Expr( current window() << Close window() );
        task_k = Expr| Function(code etc )

    Dialog:
       usr = New Window ("title", ..., ); //end New Window
/Interactive>
```

The custom **Fit Y by X** dialog box is pseudo-modal, like all JMP platform dialog boxes. The user has access to the full JMP UI. The analysis is expected to run when the user clicks **OK**. Given that the analysis code has not yet been written, creating a separate JSL script to do the analysis is probably better than adding a set of functions and expressions before the dialog box script in the same file.

Concerns and Considerations

The methods to convert user input into output entirely depends on the application goals and the expected output. However, there are several general concerns worth mentioning.

1. How will the information be transferred from the dialog box to the analysis code? Globals? Files written to disk? Passed by reference?

2. Test for valid responses and corner cases. For example, are data available? Constant response? Data source mismatch? Is a variable nominal in one file and continuous in another?
3. Know what each JMP analysis application requires.

The first concern is minor for modal dialog boxes that assign the user settings to global variables or to a global container. For analyses defined with expressions, the best recommendation is to add commands like **Show**(Char(Name Expr (myExpr))). Working with expressions can make code much easier to read. But, they make it more difficult to test or debug without statements or methods to see the commands being sent to JMP. Local versus global variables and namespaces need consideration. The practice of creating a global associative array or global list for dialog box results is a recommended practice.

The second concern is valid. However, testing might be unnecessary. The variable _yvar_clb requires numeric variables, so testing for data type is unnecessary. Data type testing is more important when reading in data. It is unnecessary to test that the user designated at least one column as a Y variable if the ColListBox option specified a minimum number.

For the third concern, most JMP platforms allow specifying columns as a list of strings. The 7_InteractiveDisplays.jsl script contains the command to create a Distribution report with mixed data type: **Distribution** (eval (_yvalList))). Yet, the **Control Chart** platform requires that each parameter gets its own Control Chart statement with attributes.

Following the directions in this section, the custom **Fit Y by X** dialog box should add the task of creating a list or global associative array, or it should make sure that _yvarList is global.

Custom Displays

Introduction

Knowing the behavior and capabilities of JMP is invaluable for efficient and effective scripting in JSL. After briefly reiterating the importance of understanding the different behaviors of active and inactive displays, the remainder of this chapter highlights some useful and fun scripting features that customize JMP graphs.

Chapter 6, "Reports and Saving Results," describes how to navigate and customize JMP reports for simple, single-variable displays and for more complex displays with Group and By variables. We call the approach in Chapter 6 customizing displays from the *top down*. This chapter presents a more flexible method of building an active display. We call this approach *bottom up*. In our context, *bottom up* means building individual displays, customizing them, and putting them together. Many scripters find that this approach implements their intentions more accurately than the top-down approach, without an egregious amount of extra overhead.

We are often thrilled and amazed by the dashboards and interactive displays scripted by our colleagues once they have learned the basics of these chapters on reports and interfaces.

Build a Custom Multivariable Display

A suitable subtitle for this section is "Syntax Versus Semantics—The Art of Building Active Displays."

Most of us learn best by example, so run the following code, which is found in the 8_Syntax_vs_Semantics.jsl script. Select the menu in each Scatterplot Matrix graph. Identical syntax produces ostensibly the same displays. However, one is an active platform display, and the other is an inactive display. The inactive display has the behavior of a journaled report. The *JMP Scripting Guide* explains this behavior:

"When display objects are created or referred to by JSL, they are freely shared references until they are copied into another display box or until you close the window and they disappear. When you plug a display object into another display tree, JMP makes a copy of it that the new box owns."

```
dt = Open("$Sample_Data\iris.jmp");
nw = New Window("Syntax vs. Semantics",   //uses $Sample_Data iris.jmp
  ob = OutlineBox("Active vs. Inactive",
   lup = LineupBox(2))
);  //end NewWindow

vL = VListBox(
sp = Scatterplot Matrix(
 Y( :Sepal length,
    :Sepal width,
    :Petal length,
    :Petal width,
    :Species )
 ));
//---identical syntax
//---different results - first display is active second is not
//---semantics
lup << append(vL);
lup << append(vL);
```

Figure 8.1 Syntax Versus Semantics: (L) Active Scatterplot (R) Inactive Scatterplot

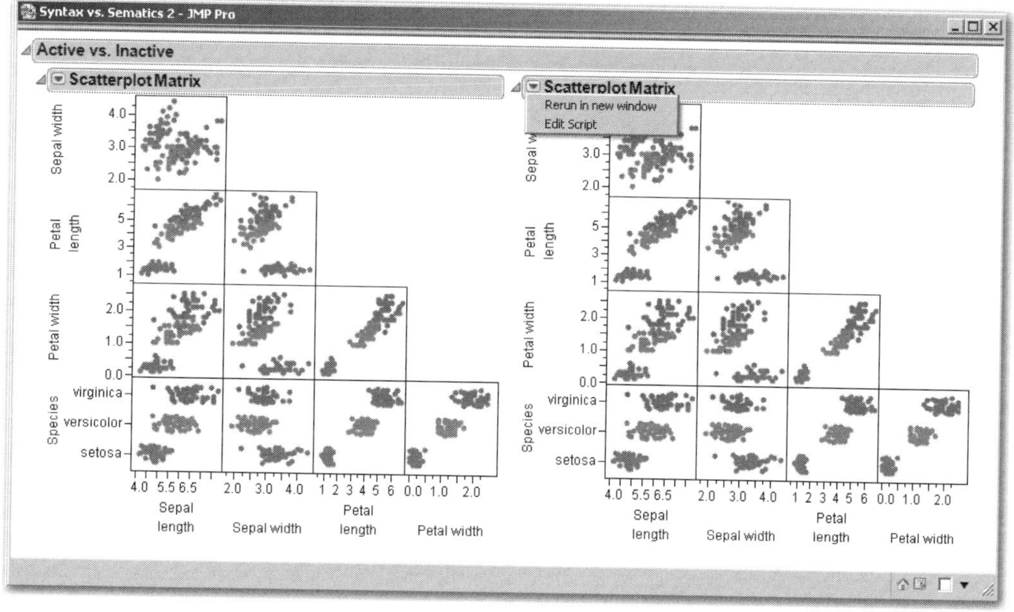

Semantics is defined as behavior or meaning in the context of usage. When a graph or platform is created, it has an internal address. Each display object must have a unique address. JMP makes a clone or inactive copy when JSL makes a request to load the same address. In this example, the second graph is a clone of vL and inactive. The 8_Syntax_vs_Sematics.jsl script includes a modification to the previous code. It creates vL1, sp1, vL2, and sp2 from identical VListBox and Scatterplot Matrix commands. When vL1 and vL2 are appended, both displays are active.

The bottom line is that you get one chance to display an object. Once displayed, referencing that object and loading it into a window creates an inactive copy. It is important to know that this copy has a new address or reference. This behavior is confusing for beginning JSL scripters, and it is important to comprehend this behavior when writing scripts to build displays. Like many things, once it is understood, it is easy to manage. However, sometimes users forget that all active displays are linked to their source tables, which means that changes to the source tables (such as row states) impact the displays.

A frequently asked question when scripters are constructing an active display is, "How do I color and mark by column Var1 in one graph, and by column Var2 in a different graph?" For an inactive display, this is straightforward. Create one graph, color by column Var1, and then journal it. Repeat this action for another graph using column Var2. Because the inactive displays are no longer linked to their source tables, changing row states does not affect graphs in the journal or any graph that is no longer linked to its source data table. Currently, there is no built-in solution to meet the scripter's original request exactly. Often, adding a **Data Filter** to the display or a few custom controls is sufficient. A perfect custom solution for this problem requires duplicate tables and scripts that save and switch row states. The key point to remember is that building an active display requires some in-depth knowledge of how JMP works. The remainder of this section discusses building active displays from the bottom up.

Build a Display from the Bottom Up

The primary motivation for learning how to build a display from the bottom up is the need to create a partitioned analysis (a By analysis) that uses multiple JMP platforms.

Figure 8.2 illustrates the difference between what JMP provides and what the user needs. The figure on the left can easily be specified with a statement such as:

```
New Window( "title", HListBox( Bivariate( By() ),
  Oneway( By() ), Overlay( By() ) ) );
```

There is no guarantee the grouping boxes will line up. In most cases, they will not. If there were 50 or more groups, the figure on the left would be difficult to use. The user layout in the figure on the right is much easier to study and interpret each group's analyses. The By group could be a species or a processing tool (from the examples in Chapter 6), or it could be an American state. In the next section, the first example uses the JMP Sample Data table SATByYear.jmp. For each state, the code plots SAT Verbal versus SAT Math scores for several years, and creates an Overlay plot of yearly results.

Figure 8.2 JMP Display: (L) HListBox of JMP Platforms (R) User Layout

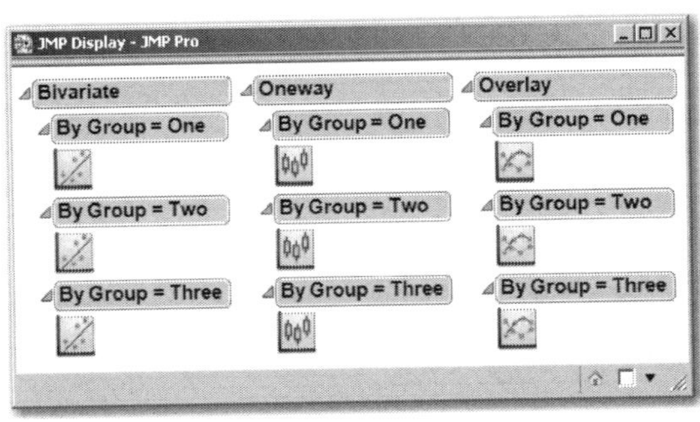

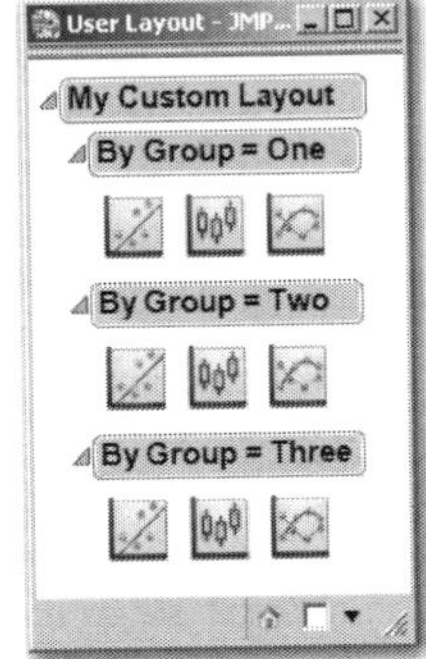

JMP Platform( ) Command

The JMP developers have provided an example of an undocumented function called **Platform( )**. The general syntax is shown in this section. The resulting display (like **ListBox**) is not drawn until it belongs to a window. Unlike **ListBox**, the specified platforms are not separated by commas. Instead, they are glued together with semicolons.

```
Platform( dataTable, platform 1; platform 2;...platform k );
```

The method to achieve the by-group analysis requires a separate table for each group. That might sound like a file-management nightmare. However, three JMP features make this a viable solution if your computer has the memory capacity to duplicate your source data. Here are snippets from the 8_CustomPlatformSubsetBy.jsl script:

```
dt = Open( "$SAMPLE_DATA\SATByYear.jmp" );
t0 = Tick Seconds();
Summarize( _b = By( :State ),    //find all States in this file
     vbmin = min( :SAT Verbal ), vbmax = max( :SAT Verbal ),
     mamin = min(:SAT Math ), mamax = max( :SAT Math )
);  //min max will be used to maintain the same scale within state
subdt = dt << Subset( By( :State ), Linked,  Invisible );

nw = New Window( "Using SubsetBy & Platform ", HListBox(
    vv = VListBox(),
    ButtonBox( "Close", nw<<Close Window() )
   ) // end HListBox
);  //end New Window

nw << OnClose( eval( subdtClose ) ) ;

For( i = 1, i <= N Items( _b ), i++,
    ymin = min( vbmin[i], mamin[i] ) - 5;
    ymax = max( vbmax[i], mamax[i] ) + 5;
    ovlExpr = parse( evalInsert( ovlStr ) );
    vTemp = VListBox(
      OutlineBox( "State = " || _b[i],
        VListBox(
          HListBox(
            Platform( subdt[i], bivExpr; ovlExpr )
        ))) // end OutlineBox
    ); // end VListBox
    wait(0);
    vv << Append( vTemp );
); // end For
```

Summarize is used to find the unique states in this file. An associative array could do the same. However, **Summarize** can also be used to find summary statistics to help with scaling the graphs. As promised in Chapter 4, “Essentials: Variables, Formats, and Expressions,” an example of **Subset(By( ))** is provided. Subsetting with a simple **By** statement creates 51 files. By linking each state's file to the main file, row states that are changed when interacting with this active report are captured in the main data table.

Figure 8.3 captures a view of the resulting window with a portion of the JMP home window and a portion of the main data table. At this point, you might want to review the discussion about invisible files. To keep the window active, the subset files must exist. The **OnClose( )** script assigned to the window nw closes the invisible files when the window is closed. The expression subdtClose is a simple For loop and is not shown. **ButtonBox** is redundant and is added as an example of a control that you might want to add to a display.

Three JMP features, **Subset** with **By**, **OnClose**, and **Platform**, enable you to create a custom, partitioned report with a little extra work.

Figure 8.3 Custom By Group Report Using Subset with By and Platform

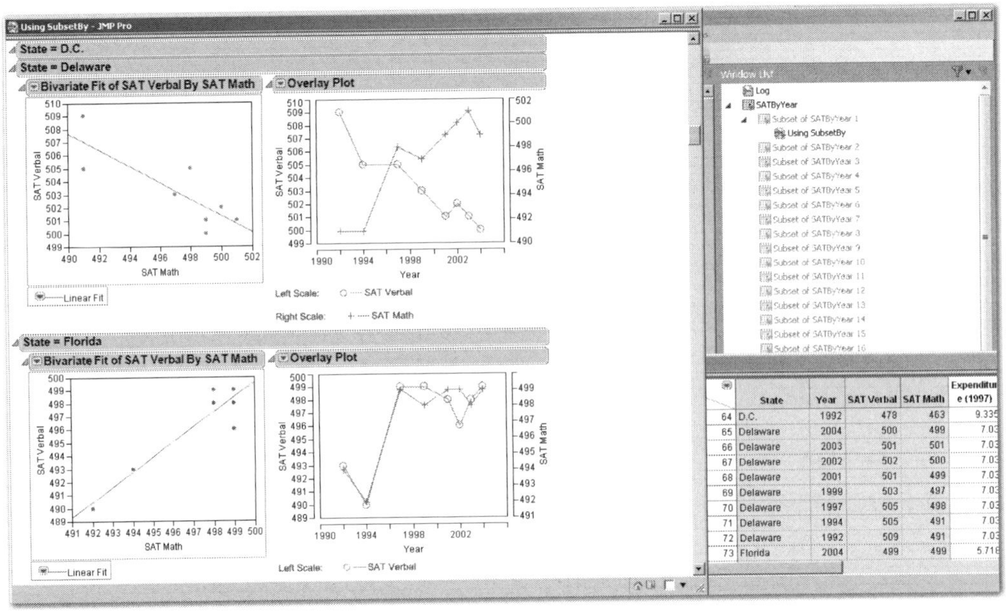

On Two Y Axes Plots

There is a time and a place for dual Y axes plots. We think that they are overused, and that they are really only appropriate when there is a fixed relationship between the axes' scales. Because the SAT scores for both Math and Verbal use the same scale, this example might not be an appropriate usage. The overlay plot for Delaware in Figure 8.3 can be easily misread. Math scores increased and Verbal scores decreased. In 2004, both scores were near 500, but visually they appear to be much different.

One common remark heard when viewing dual axes graphs is, “To which axis does this curve belong?” In Figure 8.3, if the color of each axis label matched its associated curve, and if the scales were the same, the overlay plots might be more informative. Good scaling is not free. A simple solution for this analysis is to not use two axes. Let JMP do the common scaling. The 8_CustomPlatformSubsetBy.jsl script is an example of finding the Y axes’ minimum and maximum values and coloring the labels. Figure 8.4 displays the revised analysis for Delaware. A constant scaling increment, Inc(5), was selected to calibrate the changes in scores. Wide (fewer) grid lines represent little change, and narrow (more) grid lines represent more change.

This script demonstrates two methods for making code more modular: expressions and the **Include( )** function. It uses an expression for the **Bivariate** statement. The overlay expression is created using an **Include** statement to read a macro-like string, then the **Eval Insert** function generates the expression. Both **Overlay** and **Graph Builder** have special features that might be easier to script using an expression or a string representation of explicit **Dispatch** statements.

Figure 8.4 Delaware Revised

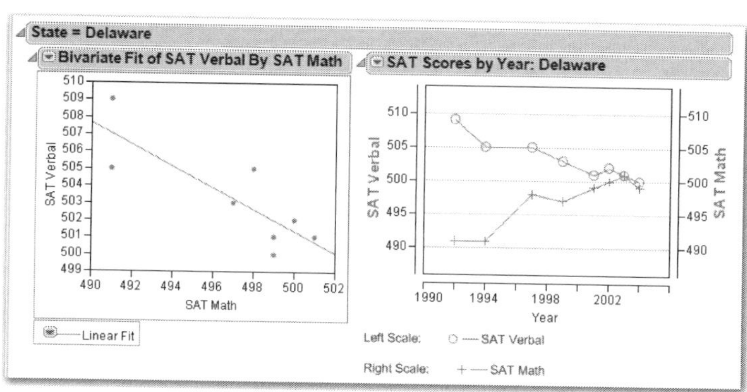

Custom By Group Analysis without Platform

This next section tackles the same task and creates a similar display. Here are the steps:

- Open the data file.
- Define all subgroups and pertinent summary information.
- Lay out the window with placeholders for graphs and analyses to be inserted.
- For each group and each analysis, create the analysis in an undisplayed container like **VListBox** or **HListBox**. Customize the display, and append or prepend to the window.
- To make the code more readable, move long, messy statements to files of expressions and strings to be included.

Creating the display yourself without **Platform( )** is just as easy, and the data is not duplicated. The following script is 8_CustomByUsingWhere.jsl:

```
dt = Open( "$SAMPLE_DATA\SATByYear.jmp" );
Summarize( _b = By( :State ) );

nw = New Window( "Using Where",
     vv = VListBox()
);

For( i = 1, i <= N Items( _b ), i++,
//---create bivariate and customize
  plt1 = VListBox( biv = Bivariate(
    Y( :SAT Verbal ),X( :SAT Math ),
    Fit Spline(0),where( :State == _b[i] ),
    Fit Line( {Report( 0 ), Line Color( {213, 72, 87} )} )
  )); // end plt1
  bivr = biv << Top Report; //the "where" text box is in the top
report
  bivr[TextBox(1)] << delete;
  bivr[OutlineBox(1)] << Set Title( "SAT Verbal vs. Math: " || _b[i]
);

//---create overlay 1 Y axis, let JMP do the scaling & customize
  plt2 = VListBox( ovl = Overlay Plot( where(:State == _b[i] ),
    X(:Year), Y( :SAT Verbal, :SAT Math ),
    Connect Points( 1 )
  )); // end plt2
  ovlr = ovl << Top Report;  //"where" TextBox is in the top report
  ovlr[TextBox(1)] << delete;
  ovlr[ OutlineBox(1) ] << Set Title( "SAT Scores by Year: " || _b[i]
);

//---Wrap the two plots in an OutlineBox and HListBox and
//   load them into the window
  vv << Append( OutlineBox( "State = " || _b[i],
        HListBox( plt1, plt2 )  )); // end Append
); // end For
```

Custom Example

The task is to create a script that can be added as a menu item and run on demand. It generates an active report window. The window should have views for overall performance, equipment performance, and individual tool trends. The parameter list is fixed, and the number of tools can vary from one to five. No user input is required to run the report.

There are numerous ways to design the report window. The script uses a simple template. Here is the strategy used to build the display:

- Lay out the window (ror_win) and outline boxes (ob_sum, ob_cmpr, and ob_trend).
- For each tool, create a report for the three parameters (T0, Time 240 degrees C, and Max ROR). Customize and append the results to the outline box named Tool Trends (ob_trend).

Keep track of issues and summary results. The display is kept active using an **HListBox** and appending the results.

- Create a tool comparison for each parameter, and append these comparisons to the outline box named Tool Comparisons (ob_cmpr).
- Accumulate summaries and write the final report to the outline box named Overall Summary (ob_sum).

Here are some snippets from the 8_CustomBottomUpReport.jsl script. The full script contains examples of building reports that can be combined, using functions, applying the platforms Oneway and Variability, and much more.

```
//create template window
ror_win = New Window( "Rate of Rise",
  LineupBox(ncol(1),
    ob_sum = OutlineBox("Overall Summary",
      obs_lub = LineupBox(ncol(1), )
    ),
    ob_cmpr = OutlineBox("Tool Comparisons",
      obc_lub =LineupBox(ncol(1), )
    ),
    ob_trend = OutlineBox("Tool Trends",
      obt_lub =LineupBox(ncol(1), )
  )
));
```

Figure 8.5 Rate of Rise: (L) Outline View of Custom Report (R) Summary and Tool Comparison

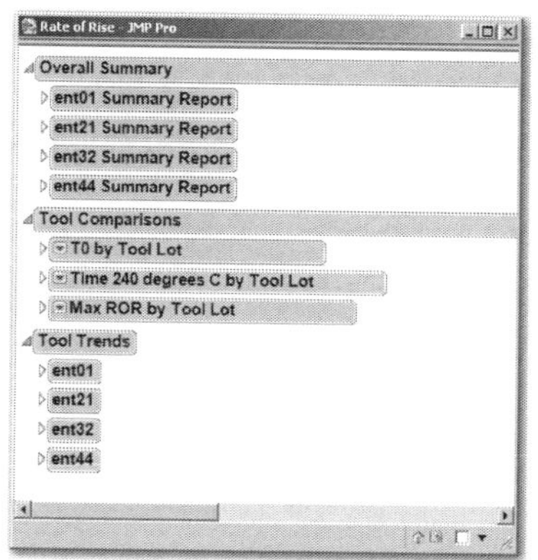

ent01 Summary Report

Tool Summary

Tool	Parameter	nLots	LotAvg	RTR	MaxWTW	MedianWTW	nOOB_Upper	nOOB_Lower
ent01	T0	23	24.9594	5.27392	2.2301	1.12694	0	0
ent01	Time 240 degrees C	23	19.5522	2.95421	2.63122	1.15326	0	0
ent01	Max ROR	23	14.8497	4.3313	21.2784	1.94207	1	0

```
//--Find names and how many tools in the recent data, this may vary
//--Associative array method to find unique tools.
ent_aa= Associative Array(column(ror_dt,"Tool"));
ent_aa << Default Value( 0 );
ent_id = ent_aa << Get Keys;  //List of unique tools
```

```
//---Create and customize Tool trends
For( i=1, i <= nitems(ent_aa), i++,
  sp = HListBox(
    eval(Substitute(NameExpr(trendExpr), Expr(_gRef), Parse("sp1"),
        Expr(_yvar), Expr(:T0)))
    );  //end HListBox  //Test//  nw=new window("test", sp);
    ModifyTrends(sp1, pLimits[1],i);
    obt_lub << Append( OutlineBox(ent_id[i], sp));

    eval(Substitute(NameExpr(trendExpr), Expr(_gRef), Parse("sp2"),
         Expr(_yvar),  Expr(:Time 240 degrees C )));
    ModifyTrends(sp2, pLimits[2],i);
    sp << Append (gg);

    eval(Substitute(NameExpr(trendExpr), Expr(_gRef), Parse("sp3"),
         Expr(_yvar), Expr(:Max ROR)));
    ModifyTrends(sp3, pLimits[3],i);
    sp << Append (gg)
);
```

trendExpr is 25 lines of a saved script. _gRef is a placeholder for the Oneway reference. _yvar is a placeholder for the response column. gg is the VListBox holding each Oneway analysis. The Oneway platform is not the usual choice for trend analyses. It was chosen to display the individual wafer results and to connect the means and the By lot summary report that JMP creates.

When writing and testing a script and customizations, select and run i=1. Run the commands within the **For** loop. Check the Log window. The last line should be DisplayBox[ ]. To take a peek, highlight and run the commented code at the end of **HListBox**. (Do not include the // comment lines.) Customizations for this report include removing the text box that uses a **Where** statement, counting the number of excursions, titles, and more.

In this example, instead of creating a single display using By(:Tool), and then navigating to each tool's report for customizations (top-down approach), a bottom-up approach is used, and each tool's display is created, customized, and appended.

Add Scripts to Graphs

The control panel for a JMP graph is its **FrameBox**. Background color, transparency, marker size, selection mode, and row legends are features that you have probably used with a right-click of the mouse. If you have not yet tried the **FrameBox** options **Customize** and **Edit**, you have seen only a fraction of the available controls.

Not all graph segments are available in JSL. However, **FrameBox** is the owner of the segments that are available.

Let's perform a quick and easy example using a frame box. Open Big Class.jmp. Run the Oneway script attached to the data table. Make sure that the display option for box plots is turned on. Right-click in the frame box. Select **Customize**, and then select **Boxplot(F)**. Change the box plot style, the color, choose to fill it, and more, if you want.

The **Edit** menu option enables the user to copy the graph contents from one graph and paste into another, copy maps and pictures to the framebox, copy and paste axis settings, and much more. The remainder of this chapter focuses on selected **FrameBox** messages and scripts that might be attached to a graph.

As with all display box scripting, the critical first step is finding and referencing the correct **FrameBox**. Once you find it, sending the correct message to it applies the customization. To see all of the scripts that are currently in a graph, right-click in the graph, and select **Customize**. A dialog box appears that lists the scripts for that graph.

Select **Help ▶ DisplayBox Scripting ▶ FrameBox** to access fun and informative sample scripts that demonstrate the properties of a **FrameBox**. You might never need to add the image of a black rhino footprint as a background in your graph (see **Help ▶ DisplayBox Scripting ▶ FrameBox ▶ Add Image**), but you do want to be able to resize your graph and add a legend. The following example uses the 8_FrameBox.jsl script and demonstrates how to reference a **FrameBox**, show available properties, and make modifications.

```
fc_dt = Open("$Sample_data\Animals.jmp");
//--make season value order to be fall, winter, spring, summer
:season << Set Property( "Value Ordering",
     {"fall", "winter", "spring", "summer"} );

vc = fc_dt << Variability Chart(
   Y( :miles ),
   X( :species, :subject, :season ),
   Process Variation( 0 ),Connect Cell Means( 1 ),
   Show Group Means( 1 ), Std Dev Chart( 0 )
);

vc_fb = report(vc)[FrameBox(1)];
Show Properties( vc_fb );  //see Log
```

vc_fb is the reference for this variability chart's **FrameBox**. The next sequence of commands change the frame size and create custom colors and symbols for each species using commands sent to **FrameBox**. Because the chart is linked to the data table (fc_dt), the data table's row states are changed as well.

```
vc_fb << FrameSize(499,180);
vc_fb << MarkerSize( 5 );
vc_fb << Marker Selection Mode( "Unselected Faded" );
```

```
fc_dt << Select Where(:species=="COYOTE");  //fc_dt<<Markers(21);
vc_fb << RowMarkers( 21 );
vc_fb << RowColors ( HLS Color( 360 / 360, 0.5, 1 ) );
fc_dt << select where(:species=="FOX");  //fc_dt<<Markers(22);
vc_fb << RowMarkers( 20 );
vc_fb << RowColors ( HLS Color( 240 / 360, 0.5, 1 ) );
fc_dt << Clear Select;
```

Marker Selection Mode is a new feature in JMP 9, and it is set to our favorite mode: **Unselected Faded**. In our dense data environment, the grayed-out, unselected points provide context without distraction. The next block of code is the answer to the frequently asked question, "How can I get JMP to create a legend and maintain the colors and markers that I set in the data table?" As long as each legend group has just one color and one marker for all members of that group, the following script (which uses the previous example) adds a legend and maintains the data table's settings:

```
//--Add Row legend to maintain table customizations
vc_fb << Row Legend( species,
Color( 0 ), Color Theme( "" ), Marker( 0 ), Marker Theme( "" ),
Continuous Scale( 0 ), Reverse Scale( 0 ), Excluded Rows( 0 ) );
```

This variability chart would benefit from a few more customizations. The connection color should be neutral, or it should match the species symbol color. **Group Means** lines for species provide information. For subject, **Group Means** lines are more of a distraction than informative, given the seasonal behavior of both species. These two properties are **CustomStreamSeg** options. Each graph has its own custom segments. The *JMP Scripting Guide* does not explicitly describe how to reference them. The command Report(vc) << Show Tree Structure reveals the IfSeg and CustomStreamSeg boxes. The best way to learn about segments is to turn on all graph features. Right-click in the frame box, and select **Customize**. (If you are scripting, type vc_fb << Customize;.) Change the colors, line width, and transparency for each segment. Save the script to the Script Window. Your customizations are found in **DispatchSeg**. For the variability graph, 8 is the **CustomStreamSeg** for connection lines. The following command turns connection lines gray:

```
DispatchSeg( CustomStreamSeg( 8 ), {Line Color( {187, 187, 187} )} )
```

No method currently exists to turn off the second group's means. So, remove the variability chart option, and add mean lines to **FrameBox** using an **Add Graphics Script** command. For each species, the custom color, the overall average, and the length of the line (xBeg and xEnd) are needed. In this example, we know that there are two species, COYOTE and FOX, with three subjects each. We know that there are four seasons, so xBeg and xEnd are 1, 12, and 12, 24, respectively. A generalized program would take the first grouping parameter, find the number of group values, and define the required information to generate the commands to handle any number of groups and subgroups. The following script makes these improvements:

```
Show(grpName, grp_xb, grp_cnt);
//line up this script with the vc window waits are added
vc_fb << Add Graphics Script(  pensize(2);
  pencolor(HLS Color( 360 / 360, 0.5, 1 ) );
  yFunction(If((x <= grp_cnt[1]), grp_xb[1]), x);
  pencolor(HLS Color( 240 / 360, 0.5, 1 ) );
  yFunction(If((x > grp_cnt[1]), grp_xb[2]), x ) );
//---------watch the lines disappear-----------------
wait(3); clear globals(grp_cnt, grp_cnt);
//click on the framebox or resize
vc_fb << FrameSize(500,180);
```

However, there is a problem with this approach. For other graph components, JMP stores the value of a variable reference. For **FrameBox**, JMP stores the reference. If you create vc_fb, journal and save it, send it to someone, and exit JMP, or if you clear global variables and then open it, the reference lines will be gone. The graphics script is intact. It's just that the values for grp_cnt and grp_xb no longer exist. The 8_FrameBoxRevised.jsl script and previous chapters include multiple examples of using expressions, and replacing expression placeholders with values.

Be cautious! Use expression substitution to replace variables with values when creating journals. If you are saving results in an image format, make sure that you save them before the global variables are changed or cleared.

More Graph Customizations

Just because you can add fancy customizations to a graph doesn't mean that you should. But sometimes, customizing a graph of rich data with color, annotations, or figures can enlighten the consumers of the graph and make the information transfer more effortless and meaningful. If customizations are used properly, these artifices become the fascia, an important part of graphic communication.

For example, run the 8_Gradient.jsl script. Here are some snippets:

```
dt=open("$sample_data\semiconductor capability.jmp");

cpp = dt<<Capability(
  Y( :NPN1, :PNP1, :PNP2, :NPN2, :PNP3 ),Goal Plot(0));
rpt = cpp << report();

//----Add A Gradient background
rpt["Capability Box Plots"][FrameBox(1)]<< Add Graphics Script(
    Gradient Function( Abs( x ) / .5, x, y, [0 3],
      Z Color( [4, 3] ),
      Transparency( .4 ));
); // end add graphics script
```

```
//---- Customize BoxPlotSeg line widths
For(i = 1; nYvar = 5; , i<= nYvar, i++,
rpt[ "Capability Box Plots" ][ FrameBox(1) ]
  << { DispatchSeg( BoxPlotSeg( i ),
     {Box Style( "Outlier" ), Shortest Half Bracket( 0 ),
      Line Width( 2 )})
    };
); //end for i
```

Figure 8.6 displays the results. You have a capability plot that compares the distributions of five parameters from the JMP Sample Data table Semiconductor Capability.jmp. Each distribution is scaled by its specification limits. This script includes statements to draw the box plot outlines with wider line widths and to add a gradient background that is drawn with a 0.4 transparency.

Figure 8.6 Capability Plot with Custom Gradient Background

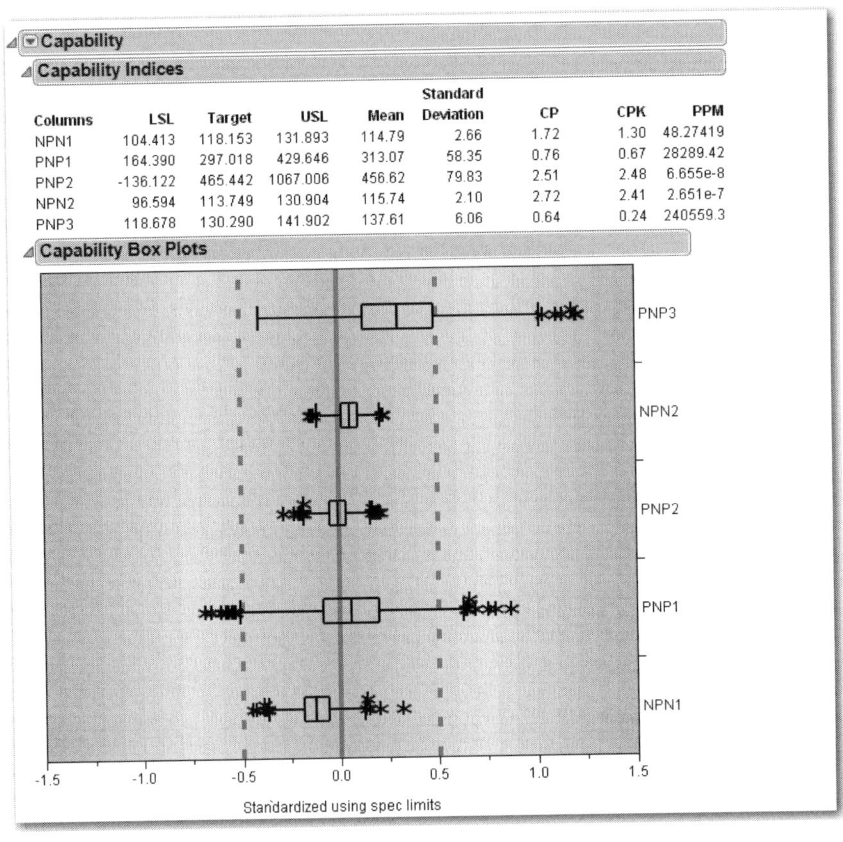

Run the script again and change the transparency. With transparency > 0.6, the gradient background is so bold, it diminishes the information depicted by the box plots. With transparency < 0.3, the background is so diffuse that the picture appears foggy.

A lot of editing and multiple trials are likely needed to create truly informative, enriched displays. This iterative task is made less arduous with the ability to highlight and run a small block of JSL statements (e.g., just the commands drawing the graph). In addition, it becomes easier if you have the ability to modify the graph settings and scripts attached to the graph interactively after a display is created. Right-click in the framebox, and select **Customize**. Figure 8.7 depicts the items in the frame box that can be edited. Select **Script** to display the gradient background script added to the frame box. Change the transparency value, and then **Apply** and repeat until you're satisfied. Or, you can remove this background completely!

Figure 8.7 Customize Graph Window

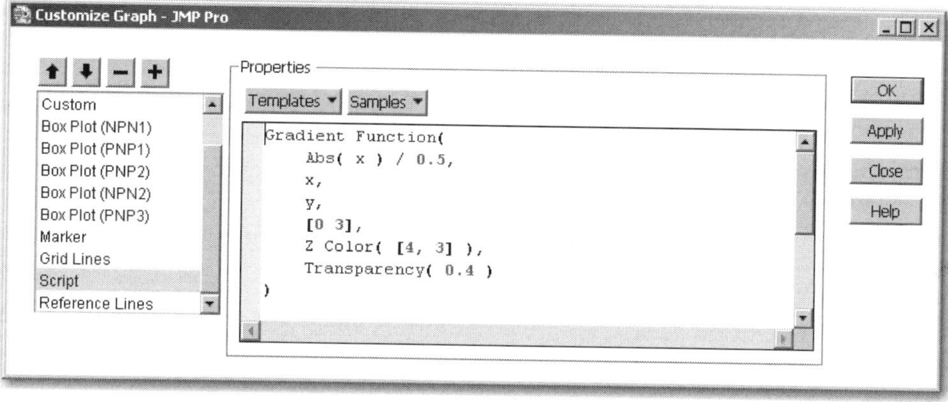

The 8_BetaSliders.jsl script uses XFunction, YFunction, and Text to draw and annotate a Beta Distribution function and to display parameters for elicitation feedback. The 8_PolygonAnnotation.jsl script uses the **Add Polygon Annotation** frame box message to draw a polygon on a graph.

Figure 8.8 Displays: (L) 8_BetaSliders.jsl (R) 8_PolygonAnnotation.jsl

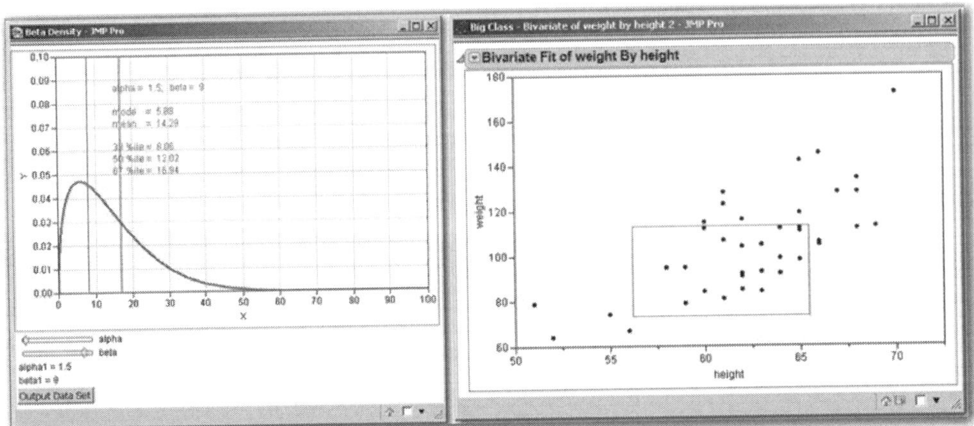

We encourage you to experiment with the different options for drawing, coloring, and annotating graphs. Apply these features judiciously to enhance the information transfer of your graphics.

Interactive Graphs

We have found interactive graphics to be very useful as training tools. There are several options in JMP to make interactive graphs. Some of these options you have seen in previous sections. For example, the 4_Expressions.jsl script uses a **SliderBox**, **GlobalBox**, and **ButtonBox** that creates Weibull distributions and generates Weibull random data. The parameters of the Weibull distributions are changed by either the **SliderBox** or **GlobalBox**, and the **ButtonBox** delays producing a table of random data until it is clicked by the user. This type of script is used as a training tool in classes when discussing the Weibull distribution. Because these types of display boxes were defined in Chapter 7, "Communicating with Users," they are not discussed again in this section. Instead, the **Handle( )**, **Mousetrap( )**, and **Drag** functions are discussed.

Handle and Mousetrap

The **Handle( )** and **Mousetrap( )** functions are used for creating interactive graphics. These functions respond to dragging or clicking. The **Handle( )** function puts a small square marker on a graph that the user drags to make changes to the graph. The **Mousetrap( )** function captures clicks on a graph, and executes the lines of code included in the function with each new click.

Here is the general syntax for the **Handle( )** function:

```
Handle(xPosition, yPosition, Script, <PostScript>)
```

The initial values of the coordinates where the marker appears on the graph need to be defined before using the **Handle( )** function. Then, the first two arguments of the **Handle( )** function define the coordinates of the marker on the graph. The Script argument in the **Handle( )** function modifies the graph when the marker is dragged. The optional PostScript argument is run after the mouse button is released.

The 8_MCADemo.jsl script uses a handle to demonstrate the impact of bias and measurement variation on the observed process variation and C_{pk}, a process capability metric. MCA is an abbreviation for measurement capability analysis. If the handle is moved up, the measurement variation increases. If the handle is moved down, it decreases. As the handle is moved right or left, the measurement bias changes. The interactive display shown in Figure 8.9 has two graphs. The top graph contains a block of text boxes of statistical indicators, the scripted handle, and two Normal distributions. The larger of the two Normal distributions represents the observed process variation, and the smaller represents the measurement variation and bias. The bottom graph contains trend data, and it is updated as the measurement variation and bias are changed.

Figure 8.9 Interactive Graph of MCA using the Handle Function

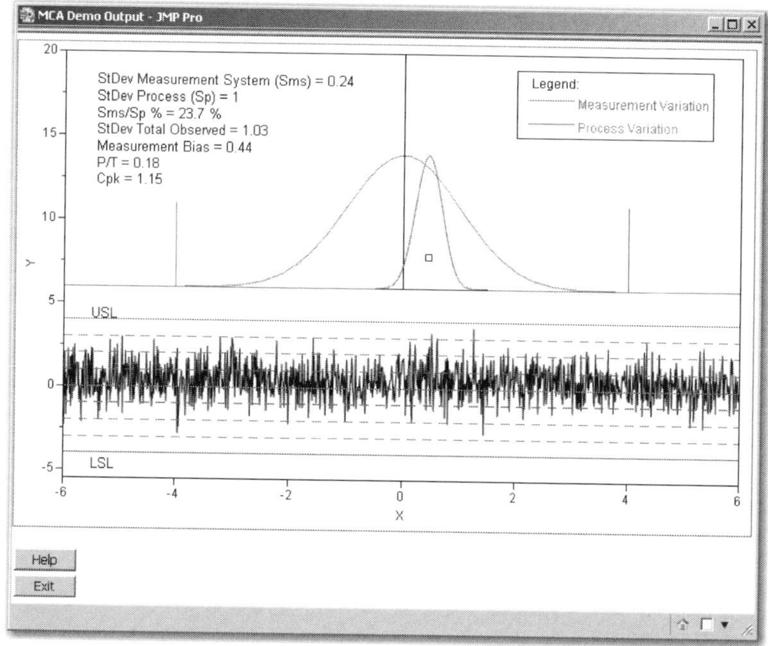

Use caution when using the **Handle( )** function. The script can be defined such that the handle runs from the mouse when trying to drag it. The following script demonstrates this problem. When the script is run, there are two handles on the graph. The first handle is at (1, 2), and the second handle is at (0.5, 1). Nothing changes until the user clicks on one of the handles. As soon as the user clicks on the first handle at (1, 2), the expression **a=2*x** is evaluated, and the handle moves. In other words, the mouse cannot catch the handle. Click or drag the second handle, and **a=2*x** is evaluated. However, the handle is drawn at (a/2, 1). The second handle is positioned at the mouse. This script is 8_DuelingHandles.jsl.

```
a = 1;
New Window( "Normal Density",
Graph Box( FrameSize( 500, 300 ),
X Scale( -10, 10 ), Y Scale( 0, 10 ), Double Buffer,
Y Function( a * x ^ 2, x );
Handle( a, 2, a = 2 * x );  //the mouse runs from this handle
Handle( a / 2, 1, a = 2 * x );  //the mouse will get this handle
));
```

The **Mousetrap( )** function captures clicks of a mouse. Here is the general syntax for the **Mousetrap( )** function:

```
Mousetrap(Script, <PostScript>)
```

When a user clicks on a graph, the **Mousetrap( )** function returns the coordinates of that click, and the coordinates can be used to update the display. The demoCorr.jsl script in the JMP Sample Scripts directory provides a good example of the **Mousetrap( )** function.

```
Open( "$Sample_Scripts\demoCorr.jsl" );
```

The script generates a scatterplot that includes a text box of the calculated correlation of the X and Y values and a normal bivariate contour. If a user clicks on the graph, a new point is added, and the correlation is updated to reflect the addition of the new point. Existing points can be dragged to new locations. This script is a great training tool when teaching correlation and the importance of plotting the data. Figure 8.10 displays the effect of a few outliers on X and Y data with a strong positive correlation.

This example is a great script to test and practice the JSL matrix skills from Chapter 5, “Lists, Matrices, and Associative Arrays.”

Figure 8.10 Correlation Demo Using the Mousetrap Function

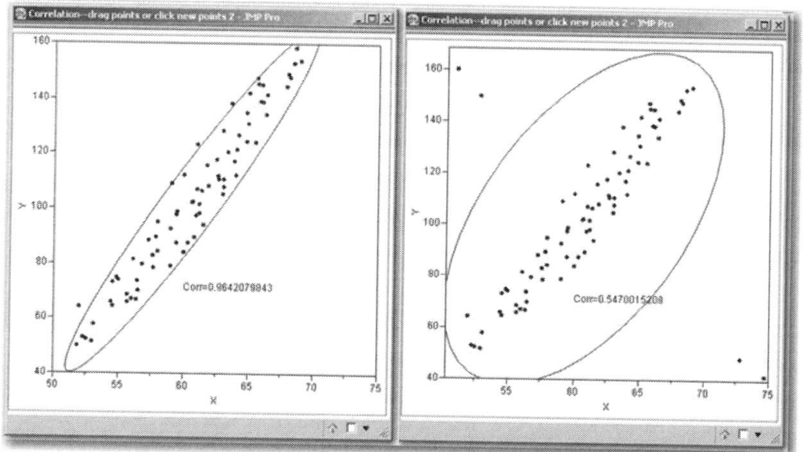

When using **Handle( )** and **Mousetrap( )**in the same script, put the **Handle( )** function before the **Mousetrap( )** function so that the handle has a chance to be moved before the **Mousetrap( )** function updates the graph.

Drag Functions

There are five **Drag** functions that enable the user to drag markers, lines, text, rectangles, and polygons. The first two arguments of a **Drag** function are matrices of the X and Y coordinates. The 8_DragFunctions.jsl script demonstrates the functionality of the **Drag** functions. It highlights the use of global variables when using the **Drag** functions (which is similar to the dueling handles script from the previous section). It also shows how all of the **Drag** functions are updated when an item is dragged if the same global variables are being used for the X and Y coordinates. Figure 8.11 shows the initial window before dragging any items.

```
New Window( "Drag Functions",
  first_x = [5 25 42],
  second_x = [55 75 92],
  first_y = [7 35 4],
  second_y = [57 85 54];
  Graph Box(
    Frame Size( 400, 400 ),
    Drag Line( first_x, second_y );
    Drag Marker( second_x, second_y );
    Drag Polygon( first_x, first_y );
    Drag Text( second_x, first_y, "text" );
    Line( second_x, first_y ))
);
```

Figure 8.11 Drag Functions Window Including Line, Marker, Text, and Polygon

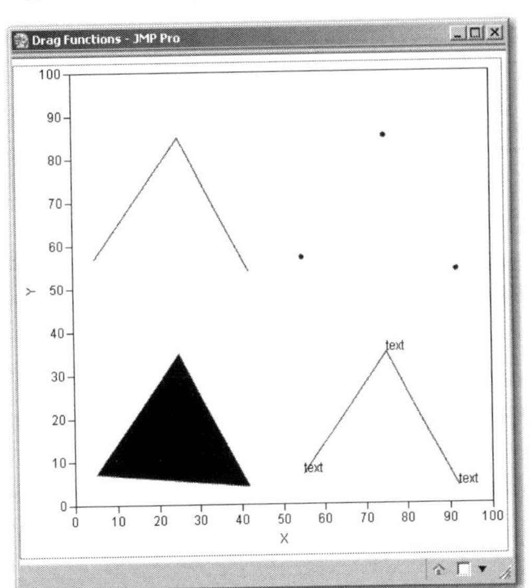

Numerous JMP platforms have built-in interactive controls. Think about the bivariate spline smoother, the data filter range selector, custom profiler sliders, bubble plots, and prediction profiler drag line controls.

Earlier in this chapter, the JMP example for Add Image, which adds a black rhino footprint to a graph, was mentioned. The ability to add a photo image of a footprint or an aerial view of landscape to a graph, and then to add interactive controls like the **Mousetrap( )** function to pinpoint subtle features, enables you to record subtle features in complex displays.

A Few Items to Note:

- A **Mousetrap( )** script can be attached to the **FrameBox** of any JMP graph. **Mousetrap( )** does not capture the clicks on a graph shape or a graph point. It will capture clicks outside of those graph features. It also returns position information for an image added to the **FrameBox**.
- JMP can read graphics images and convert them to a two dimensional array of RGB (red green blue) values. Digital signal processing and other analyses can be applied using JMP matrix functions.
- When adding custom graphics to a **FrameBox** or **PictureBox**, check whether it is drawing in pixels, which are relative to the size of the graph, or x-y coordinates, which scale with the data. An example of this is included in the 8_PolygonAnnotation.jsl script.

Writing Flexible Code

Introduction

The word "flexible" is often used in the context of an object being able to bend without breaking, or something or someone being able to adapt to new circumstances (being able to change or be changed).

This chapter covers JSL objects and methods and our recommendations to help you write flexible scripts that run without breaking and are easily maintained or extended.

The first section, "Code for the Task," provides software quality characteristics that define flexible code. Subsequent sections provide descriptions and examples of JSL objects and methods that facilitate these characteristics.

Code for the Task

Most first scripts are written to eliminate a time-intensive repetitive task. Consider the task to create a well-defined report in an interactive analysis window. You generate this report weekly. You might be able to create your script in minutes using JMP-written JSL statements.

However, it is not uncommon for the scope and usage of any script to expand quickly. When it becomes a shared program that others use in multiple situations and possibly multiple environments, it is time to extend or rewrite the code to make it flexible. Table 9.1 is a list of quality characteristics that are often presented in computer science courses or recommended for software projects.

Table 9.1 List of Quality Characteristics for Software Projects

Characteristic	Description
Compatibility	The ability to operate with different products.
Extensibility	New features can be added without major changes to the architecture.
Maintainability	The code is well documented, organized, and easy to debug.
Modularity	Programming tasks are broken into independent components that can be written and tested separately, and then assembled.
Packaging	The program is easy to install, script requirements and dependencies are documented, support ownership is identified, and a method to deliver upgrades is defined.
Reusability	Code can be reused in new applications with little or no modification.
Robustness	The program handles invalid input and stress. This includes stresses like many variables, observations, objects, and calculations. The script fails gracefully.
Security	The program cannot be destroyed or contaminated.
Usability	The user interface (UI) is intuitive and user friendly, and the program provides functionality for a wide range of situations.

Each quality characteristic requires extra coding. We use the phrase "code for the task" to imply the coding effort should be proportional to the script's requirements and use. For example, adding extensive help and unbreakable security to a single-user weekly script is probably wasteful.

Code for the task is our way of saying write a script that provides quality and value, but don't be wasteful.

Compatibility

The last two sections of this chapter present JSL scripts to call other programs. Other than the commands to call other programs, JSL executes solely within the JMP environment. JMP provides several utility functions to check the environment of the user. The 9_Compatibility.jsl script includes examples of these utility functions. Four useful functions to remember are **Host Is( )**, **JMP Version( )**, **JMP Product Name( )**, and **Get Platform Preferences( )**.

```
versionInfo = Eval List( {Trim( JMP Version() ),
     JMP Product Name()} );

osInfo = Eval List( {Host Is( Mac ), Host Is( Windows ),
       Host Is( Bits32 ), Host Is( Bits64 )} );

usrBivOpt = Arg( Get Platform Preferences( Bivariate ) );

Show( versionInfo, osInfo, usrBivOpt );
```

Extensibility

Learning to write extensible code comes with experience. Even then, there comes a point when code might benefit from or require a redesign. For example, large applications running in JMP 8 might benefit from a redesign using **Namespace**, which was introduced in JMP 9. For novice scripters, the simplest method to make code extensible is to replace hardcoded names and values (including **For** loop values) with reference variables. You can either define the reference variables at the beginning of your script or you can prompt for them. Suppose the data table has the format of Semiconductor Capability.jmp, which has four columns of identifiers. The remaining columns in the data table are process parameters to be analyzed. If the following code is used to define the columns to be analyzed, it is easy to change the number of column identifiers and the columns to be analyzed:

```
for( i = 5, i <= 132, i++, v = column( i );  ::analyze( v ) );

//more extensible: less editing if number of ID/process cols change

idList = {"lot_id", "wafer", "Wafer ID in lot ID", "SITE"};
for( i = nitems( idList )+1, i <= ncol( semi_dt ), i++,
  v = column( i );
  ::analyze( v )
);  //end for
```

Maintainability

Good naming practices, proper scoping, keeping algorithms simple (or at least well documented), and eliminating cookie-cutter code are important for maintainability. Cookie-cutter code refers to the practice of creating similar commands with copy and paste. Creating functions and expressions that can be reused within the script is preferable to writing cookie-cutter code. For example, in the previous example, the **::analyze( )** function can be changed and tested in one place, instead of finding and fixing all occurrences in the program.

Modularity

JSL is not a compiled language with linked libraries. Modularity is achieved with expressions, functions, and **Include( )**. A common practice with large applications is to store a series of functions in a separate file or in several separate files, and then use **Include** statements at the beginning of the main script. Suppose you have three functions that are used often. getCompanyLogo(width) returns a PNG file to be used in a journal file. robustScaling(value_vector) returns an expression to scale the axes to the intrinsic distribution of the data. And, getCompanyShift(datetime_value) returns a character string for a date and time argument. Suppose these three functions consist of 160 lines of code and comments. The code for these functions can be stored in one file and opened with a single **Include** statement or they can be stored as separate files and require three **Include** statements.

Packaging

JMP 9 has several new functions that enable JSL to check for operating system and environment dependencies. See the 9_Compatibility.jsl script for examples. Also new in JMP 9 is the add-in feature, which makes it easy to install an application that requires several files.

Reusability

Reusability is typically a by-product of modularity and maintainability.

Robustness

Robust code includes **Try( )**, **Throw( )**, and type-checking (such as **IsMatrix**, **IsExpr**, etc.) statements. The next section in this chapter, "Capture Errors," includes examples and recommendations for making code robust.

Security

We share our scripts with our colleagues because we believe reading open-source code (JSL or any language) is the best way to learn and share ideas. However, for support reasons, you might want to encrypt the code installed with an add-in, and store the unencrypted code on a read-only shared drive. From the JMP menu, selecting **Edit ▶ Encrypt Script** provides menu options to attach a password or separate passwords for running and reading code.

Usability

Chapter 7, "Communicating with Users," discusses characteristics and methods for creating a user-friendly UI. Being able to use the same code for different scenarios earns it high marks in usability. **Fit Y by X** is an excellent example of code that is user-friendly and can be used for multiple Y values and multiple X values with varying modeling types (such as nominal versus continuous, continuous versus continuous, etc.).

The remainder of this chapter documents methods for making your code flexible.

Capture Errors

Robust code detects and handles anticipated errors. Usually, the amount of error capture code is proportional to the risk and the severity of an error. For example, at least 70–80% of code written for a robotic surgical arm consists of commands to capture errors and gracefully and safely handle them.

In contrast, if you are writing a script for yourself or a colleague to generate a simple report, you might have no error-handling code whatsoever.

We emphatically recommend that you incorporate some error-handling and exception-handling code if you are writing an application that will be used by many people. The first step is anticipating errors.

Anticipate Input Errors—What If?

We all learn by example and experience. When reading others' code, look for examples of error checking. If you are impressed with the level of error capturing, the programmer is probably an experienced scripter and is knowledgeable of the data and environment. It is probably not the first version of the script either. Here is a starter list for anticipating input errors. What if any of the following were true?

- The script uses **Current Data Table( )** and currently no data table is open.
- The script expects a specific column name and the target data table does not have that column.
- The expected column is found, but it is the wrong data type.
- A file exists and is opened (or the SQL query runs), and the resulting data table has zero rows.
- An argument for a function is the wrong data type. For example, getCompanyShift(datetime_value) expects a numeric argument. The user's table has a column that looks like a datetime numeric value, but it has the character representation of a datetime.

Checking for these types of errors can be described as checking for existence and type-checking. This is the first level of capturing errors. Here are two snippets from the 9_CapturingInputErrors.jsl script. This script includes examples and recommendations when looking for date and time columns and more.

```
//N Table() returns the number of currently opened tables
If( N Table() == 0,
  lj_dt = Open(),
  lj_dt = Current Data Table()
);
```

```
//Check that an expected column exists
lj_dt = Open( "$Sample_Data\Chips R Us.jmp" );
colStrList = lj_dt << Get Column Names( string );
If( N Row( Loc( colStrList, "Time" ) ) == 0,
  //prompt for which column to use for time or abort
promptExpr
);
```

Suppose you are writing a script to open multiple text files. You want to concatenate the resulting tables into one large file. If one or more of the files has the proper column headings but no rows, JMP is able to successfully execute the concatenate task. However, you might not get the expected outcome. Code to check for this problem is called error handling. Here is an example of error-handling code: If (NRow(new_dt) == 0 , Close(new_dt, NoSave)). In addition, from the JMP menu, you can select **File ▶ Preferences ▶ Text Data Files ▶ Import Settings**. Locate the checkbox **Treat empty columns as numeric**. This preference setting can be useful for preventing unwanted results.

The last What If?, checking that a character column named datetime is a proper datetime string, can be a daunting task in many languages. Many languages use regular expressions. One regular expression example that received accolades on the Web was more than 300 characters long. It parsed only one datetime format, and it required leading zeros for numeric values less than 10. If the dates in your data table are post 1904, JMP can convert a column of mixed datetime formats to numeric date values, and no leading zeros are required. The 9_CapturingInputErrors.jsl script contains this mixed-format example. For a simple test, use any valid date or datetime format as an argument for the **Parse Date( )** function. This is so much nicer than a regular expression.

```
xx = Parse Date("JUN162010 2:59:13");
If ( Try(xx = Parse Date("JUN312010 2:59:13")); 1,
  Throw("Not a valid date")  );
```

Exception Handling with Try and Throw Functions

Try( ) and **Throw( )** are described in JMP online Help by selecting **JSL Functions ▶ Utility**. **Try( )** has two arguments of type expression: **Try(** expr1**,** expr2 **)**. Recall that an expression can be an entire script. JMP tests expr1, and if any errors (exceptions) occur it runs expr2. If no error or exception occurs, expr1 is run. Typically, expr2 contains commands to document and handle the error or exception.

Throw( ) can have one argument, which is a text string. When the throw command is executed, the text string is written to the Log window, and program execution is halted. The following code is in the 9_TryThrow.jsl script. **Matrix Box(** xx **)** throws an exception because xx=25 is not a matrix. The Big Class data table is not opened, and the throw command halts execution.

```
xx=25;
//task #1: same as incl1.jsl
Try( New Window( "Example",
   mb = Matrix Box( xx ));
   1 ,   //returns 1 if successful
  err_str = "xx is not a matrix..aborting";
  Caption(err_str);wait(3); Caption(Remove);
  Throw(err_str) ;
);
//task #2: same as incl2.jsl
open("$Sample_data\Big Class.jmp");
```

Alternately, **IsMatrix(** xx **)**, **IsList( )**, **IsNumber( )**, or **Type( )** functions can be used to check xx and handle the exception instead of halting execution. If a **Throw( )** is executed within an included script, only that script's execution is halted. If the two tasks in the previous script were written as two **Include** files, the second task, which opens Big Class, would occur. **Throw( )** halts the execution of incl1.jsl, not the entire program. However, by coding the **Try( )** function to return a 1 if successful, the value of inc1 reveals the success or failure of the first **Include** command. The 9_Try ThrowInclude.jsl script #3 demonstrates how to use the return flag of an include file to control program execution.

```
//xx = [1 2 3 , 4 5 6];
xx = 25;
inc1 = Include( "$JSL_Companion\incl1.jsl" );
inc2 = Include( "$JSL_Companion\incl2.jsl" );
Show( inc1, inc2 );
```

JMP handles many scenarios gracefully: an empty is returned instead of an exception. For example, yval=Log(-1); returns yval empty. Sqrt("4") throws an exception and returns an empty.

Robustness is an important quality of flexible code. In summary, anticipate errors, check for existence, and check for type. **Throw( )** halts the execution of the calling script, if needed. Use **Try( )** sparingly for limited lines of code. **Try( )** captures errors, and the Log window does not document the error's location. See the 9_TryThrowInclude.jsl script for details.

Use Namespaces

Good naming practices, proper scoping, and using namespaces are techniques for maintainability and extensibility. Suppose you have the task to add a new analysis method to an existing program. Assume that the program is similar to the Bivariate platform, and you are asked to add a Lowess fit to the list of fitting methods.

Code for a rich program like the Bivariate platform could easily have thousands of lines of commands. A flexible program would have two or more files (modules) for each method. For example, there might be one file or function to perform the Lowess calculations, and two or more files or functions to manage the Lowess report window features. Also, it is likely that the program would maintain an associative array of functions and an associative array of controls. The bottom line is that whether the inherited program is one huge monolithic coding nightmare or a well-organized set of smaller scripts, scoping and namespaces will avoid name collisions. The monolithic coding nightmare requires strict scoping and unique namespaces. With modular code, you can get by with simple **Default Local** and **Names Default to Here(1)** commands and less strict scoping.

A Lowess fit called a *kernel smoother* is an experimental feature of the Bivariate platform since JMP 7. If you hold down the SHIFT key and select a platform menu, you might see older methods or experimental methods. Simpler tasks than adding a Lowess fit are used to discuss namespaces.

Recall that a namespace is a collection of names and their corresponding values. Review the section, "Scope Variables," in Chapter 4, "Essentials: Variables, Formats, and Expressions," and the script 4_MoreScopingNamespaces.jsl for namespace basics.

Namespaces are new to JMP 9. User-defined namespaces share many of the same messages as an associative array, such as **<< Insert(varname, varvalue), << Get Keys, << Get Contents**, and **<< Get Values**. The values of a namespace are stored in an assignment list. Here is the basic syntax:

```
nsref = New Namespace( nsname, { name1 = expr1, name2 = expr2, … } );

Names Default To Here( 1 );
nsOne = New Namespace ( "nsOne", {x=11, y=12, z=13});
show( nsOne, nsOne << Get Contents, nsOne:x, nsOne:y);
show( nsOne["y"], "nsOne":z);
Delete Symbols();
Clear Globals();
Show Namespaces(); //nsOne still exists
```

There are several methods to reference an item in a namespace. For example, nsOne:y, nsOne["y"], and "nsOne":y. Refer to the *JMP Scripting Guide* or JMP online Help by selecting **Object Scripting ▶ Namespace** for all syntax options. **Delete Symbols** and **Clear Globals** clear the assigned value to nsOne (nsref), but they do not delete **Namespace** ("nsOne"). The following code demonstrates how to delete a namespace by name or by reference (if it exists).

```
//JMP9 a namespace can only be deleted by reference << delete;
Namespace( "nsOne") << delete;
//----- or use
// nsOne << delete;
// prior to deleting symbols and globals
Show Namespaces(); //nsOne no longer exists
```

Namespaces Are Global

The 9_Namespace.jsl script creates a namespace in a Local function. It also creates a JSL script that is run with the statement **Include**(jslfile_path, <<New Context). **New Context** is an option that tells JMP to run the included file in its own anonymous namespace, independent of the calling (parent) script.

```
// Namespaces are global
testit = Function({}, {Default Local},
   nsTwo = New Namespace( "nsTwo",
   {x=21, y=22, z=23});
   1
);
xx=testit;  //run the function
//run one line at a time
Show Namespaces();
//namespace nsTwo exists is not Local
Show(nsTwo);
//the reference is Local, does not exist
```

nsref Versus nsname

These two examples from the 9_Namespace.jsl script highlight the fact that nsref, a namespace reference, can be a source of confusion. If you are creating a user-defined namespace, we suggest you keep the namespace name (nsname) intuitive and simple. We recommend using the same name for nsref and nsname.

Expressions in Namespaces

It is a trivial task to insert an item whose value is an expression or a function. However, when referencing a namespace item, an implicit evaluation occurs. The following code creates a namespace with functions and expressions. The **Show** commands demonstrate this evaluation.

```
app = New Namespace("app");
  app:greetings =Function({}, "hello world");
  app:time = Function({}, Today());
  app:memory = Function({}, Get Memory Usage());
  app:fcn = Function({}, NameExpr(Expr(x + 3)) );
  app << Insert("xpr", Name Expr( x + 3 ));
  app["nmexpr"] = Name Expr( Expr(x + 3) );
  app:nmexpr_2 = Name Expr( Name Expr(x + 3) );
  app:strexpr = Parse( Char( Expr(x+3)));
Show( Namespace( "app" ));

show(app:xpr);
y = app:xpr;  //throws an error if x does not exist
```

The namespace items app:fcn and app:nmexpr_2 return an unevaluated expression. The real benefits of namespaces are the abilities to lock, unlock, and delete all variables in a namespace with a single command; the **New Context** message for include files; and the **Names Default to Here( )** function.

For graphing, window and box namespaces simplify the tasks of sharing interactive controls and making them independent. There are six fun and informative graphics examples in the 9_Namespace.jsl script.

Deploy JMP Scripts

Once a script is written, there are several options for deploying it so that it is easily accessible to many users. Arguably, the most common methods are to attach it to a data table, add it to a menu or toolbar, or, beginning in JMP 9, create an add-in. For a script that is run on the same data table (perhaps with the data table being updated periodically), the best option is to attach the script to the data table. For scripts that are commonly used either by an individual JMP user or a production script that is used by many, we recommend that you create an add-in. In JMP 9, using add-ins makes it very easy to share simple and complex scripts with many users.

Attach and Run Scripts from a Data Table

Chapter 4 discusses adding scripts to data tables, and then retrieving and executing these scripts. For thoroughness, a brief discussion of deploying a JMP script from a data table is included in this section. Consider the simple example of adding a script to the sample data table automess.jmp. The 9_DeployingScripts.jsl script fixes the data table, creates a map of the United States, and colors it by the number of auto thefts.

Here is a snippet of the script that shows how, after correcting the error in the original data table, the script is added to the data table and run. See the entire script for details.

```
//Add script to the data table
autocorrected_dt << New Table Property( "Auto Theft by State",
Graph Builder(
Variables( Color( :Auto theft ), Shape( :State ) ),
Elements( Map Shapes( Legend( 2 ), Summary Statistic( "Mean" ),
Show Missing Shapes( 0 )))));
//Assign script to a variable and show variable
auto_theft = autocorrected_dt << Get Property( "Auto Theft by State"
);
//Execute script
auto_theft;
```

JMP Add-Ins

Sometimes JMP adds a feature or platform that, because of its power and usefulness, completely changes how you approach a task. Add-ins will likely change your approach to packaging and deploying scripts. An add-in enables you to bundle scripts, data tables, pictures, and other types of files that are needed by your application into a single file. The user installs the add-in by simply opening this bundled file and clicking **Install**. It makes installing new applications very simple. Add-ins can be sent to even the most novice JMP user with confidence that the installation will be successful.

The *JMP Scripting Guide* has an excellent section on how to create add-ins. We recommend that you follow it when creating your first add-in. The following is an overview of the critical steps. Creating an add-in is not as complicated as you might think.

1. Register the add-in (as a .def file).
2. Define how the add-in will be deployed (in a .jmpcust file).
3. Bundle the files and deploy (as a .jmpaddin file).
4. Test the add-in.

Of course, the steps assume that an application is written, that it runs free of errors, and that it is ready for deployment.

Figure 9.1 is a list of files used to build an add-in called ColorCells. The add-in launches a script that colors cells based on a selection using the data filter. Try installing the add-in by opening the ColorCells.jmpaddin file. You will notice that it is added to the **Add-ins** menu and toolbar. Here are descriptions of the files involved with this add-in:

- 9_4_ColorCells.jsl—This file contains the script that is run when the add-in is launched.
- addin.def—This is the registration file.
- addin.jmpcust—This is an XML file that contains deployment information for menus and toolbars.
- addin_Backup.jmpcust—JMP adds backup files to your directory. These files do not need to be bundled with the add-in.
- ColorCells.jmpaddin—This is the add-in file that was created and can be sent to users. Users need to open the file in JMP and click Install. (Change the file extension to .zip to read the files required for the add-in.)
- ColorCells.png—This is a picture to be used on the menu and toolbar for the application.

Figure 9.1 Files for ColorCells Add-In

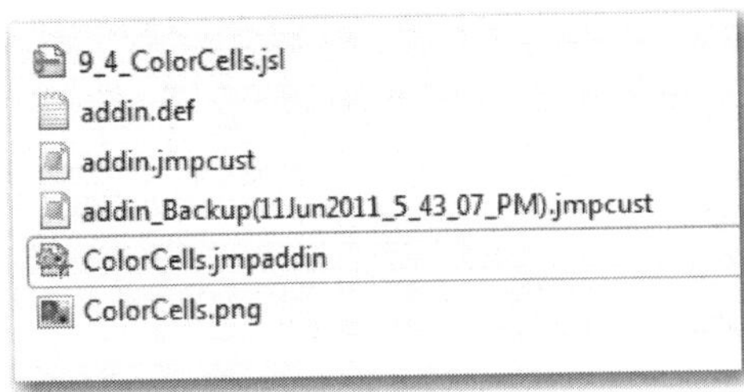

The directory for building an add-in needs to include all of the required files for the application (scripts, data tables, etc.), the registration file, and the customization file. The script in this example is fairly simple and has only one script file. If the script uses an **Include** function, the scripts that are included need to be in this directory also. For more on the **Include** function, see the section, "Use Namespaces."

Once the directory contains all of the necessary files, simply zip all of these files into one file, and change the file extension from .zip to .jmpaddin. The .jmpaddin file can then be opened in JMP and installed. To see how this works, change the extension on the add-in ColorCells.jmpaddin so that the file is called ColorCells.zip. Open the zip file. It contains the files from the list with the exception of the backup file (because it is not needed).

Open the addin.def file in a text editor. You will see that registering an add-in needs only a few lines. The first two options, id and name, are required. The id is the identifier that JMP uses for the add-in. Identifiers need to be unique within a JMP session, so it is important to make the identifiers somewhat specific to your application and organization. The name is the user-friendly name of the add-in that is seen by users. The third line is optional, and it sets the add-in to load when JMP is deployed. There are naming tips for id and other options in the *JMP Scripting Guide*.

```
id = "com.jslcompanion.colorcells"
name = "Color Cells"
autoload=1
```

Open the addin.jmpcust file in a text editor. It is a bit intimidating and a little overwhelming. Thankfully, JMP actually writes this file. All that you need to do is copy it to your directory after defining how the add-in will be deployed. This file is written by JMP when you customize a menu and toolbar for your add-in. See *Using JMP* to determine where the customization file is stored.

Add-ins are an excellent way of deploying scripts, whether the script is for personal use or for masses of people. There are several examples of JMP add-ins, including an Add-in Builder, on the JMP Web site www.jmp.com.

Menus and Toolbars

A section in *Using JMP* provides information about personalizing JMP. It includes how to add new menu items and toolbars. It is highly recommended that you read this section to learn about customizing JMP menus or toolbars and to get help with creating the .jmpcust file.

JSL Functions

Functions let you modularize parts of code that are run repeatedly. This relieves the scripter of the burden of explicitly writing the code every time it is needed. What actually goes on in the function is usually not a primary concern of the user. More important to the user is that the function transforms the inputs into useful outputs. Modularization is the main reason for using functions in JSL. Examples earlier in this book include a function to convert text to title case (also called mixed case) in the 3_ManipulatingRows.jsl script. There is a function to draw a sample swatch of color in the 5_AssociativeArrays_Dictionary.jsl script, and functions to simplify code, such as moving column names from one ColListBox to another or toggling a button or checkbox setting in several scripts in Chapter 7.

Most industries have standard analyses and complex computations. For example, in semiconductor manufacturing, a scripter might want to modularize activities like estimating the temperature of a gas at a specific volume and pressure, transforming Cartesian coordinates to polar, or calculating the coefficient of variation of an effective process cycle time. Each of these tasks involves mathematical detail, and each task could be considered basic to the profession with the possibility of being used frequently. The tight control over the input-gives-output model enabled by functions can be the key to getting reliable results. Functions can be an important basis for an advanced scripting project.

Here's how a function is specified:

```
Name = Function({arguments}, <{local variables}>, script);
```

The scripter provides the list of input arguments, an optional list of variables that are local to the function, and any valid JSL script. The function returns the result of the script based on the specified arguments. With the exception of the last example, the scripts for this section are in 9_FunctionScripts.jsl.

Suppose that you want to create a function that takes the square root of an input but tolerates negative arguments, and returns imaginary numbers rather than errors. You first specify the arguments in a list with curly braces { }, and then state the expression directly.

```
iRoot = function( {x},
  if(x>0,
     char(sqrt(x)),
     trim(char(sqrt(abs(x))))||"i")
);
a = iRoot(9);    // result in a is "3"
b = iRoot(-9);   // result in b is "3i"
```

Functions are stored in global variables, but they can be declared as local to another function or any script so that they do not affect the global symbol space. Use **Default Local** as the second argument of the **Function** definition to ensure all unscoped variables are local values. This is practical for temporary variables and essential for recursive functions, which need to keep the values of the local variables separate at each level of evaluation.

```
Name = Function({arguments}, {Default Local}, script);
```

The use of **Default Local** localizes all the names that have the following characteristics:

- Are not scoped as global variables (for example, ::name).
- Are not scoped as data table column names (for example, :name).
- Occur without parentheses after them (for example, are not of the form name(...)).

For example, the following trivial function calculates the polar coordinate r from the Cartesian input by a root sum of squares approach. It uses a local temporary variable called, creatively, temp.

```
polar3r = Function({a, b, c}, {Default Local},
   temp=a^2+b^2;
   sqrt(temp+c^2)
);
r=polar3r(3, 7, 31);   //log returns 31.9217793990247
```

The temp variable is not a global variable. However, if it is already a global variable, it remains untouched by evaluating the function. If a scripter uses an expression initially as a local variable, then uses it as a global variable, JSL changes the context.

The **Recurse** function makes a recursive call of a function. For a classic example, you can create a function to calculate factorials. The factorial of a nonnegative integer *n* is the product of all positive integers less than or equal to *n*.

```
Myownfactorial = function( {x},if (x==1, 1, x*recurse(x-1) ) );
myownfactorial( 31 );   // log returns 8.22283865417792e+33
```

Of course, JMP has an innate **Factorial( )** function, which would be easier and more efficient to call in a script. You can define recursive calculations without using **Recurse**, but using **Recurse** avoids name conflicts when a local variable has the same name as the function, and a scripter can use **Recurse** even if the function itself has not been named.

Functions can be used to manage row states, create displays, draw with pixels, manipulate nonnumeric data, and even to perform mathematics (as you might expect).

This example uses a function and an associative array to draw swatches of different crayon colors. The script is in 9_CrayonSwatches.jsl.

```
//Define a function to reference crayon colors from an Associative
Array
see_color = Function( {rgbName},{Default Local}, drawit = Expr(
New Window( rgbName, Graph Box( framesize( 150, 150 ), Pen Color(
_rgb_ );
Fill Color( _rgb_ );Rect( 10, 90, 90, 10, 1 );)));
swatch = Substitute( Name Expr( drawit ), Expr( _rgb_ ),
::rgb_AA[rgbName] );Show( swatch ); swatch;);

//Create an Associative Array using a list of key-value pairs of
crayon colors
rgb_AA = Associative Array(
{{"Atomic Tangerine", [255,164,116]}, {"Caribbean Green",
[28,211,162]}, {"Electric Lime",[206,255,29]}, {"Granny Smith Apple",
[168,228,160]}, {"Macaroni and Cheese", [255,189,136]}, {"Navy Blue",
[25,116,210]}, {"Orchid", [230,168,215]}, {"Piggy Pink",
[253,221,230]}, {"Razzmatazz", [227,37,107]}, {"Sunset Orange",
[253,94,83]}, {"Tropical Rain Forest", [23,128,109]}, {"Wild Blue
Yonder", [162,173,208]}});

//Draw swatches of specified crayon colors
see_color( "Tropical Rain Forest");
see_color( "Caribbean Green" );
see_color( "Granny Smith Apple" );
```

Parse Strings and Expressions

With the explosion of data from what seems to be all directions, the format of data files varies dramatically. It is almost certain that the data need to be cleansed, and this generally involves data as strings. Being able to extract and match pieces of information from strings can be critical to obtaining the most information from your data. Doing this efficiently is necessary with the large amounts of data that are generated daily.

Character Functions

Most of the time JMP character functions are sufficient when parsing strings or cleansing data files.

There are a variety of character functions available in JMP, and these can all be scripted. The indexes available in JMP online Help are an excellent reference. In the **JSL Functions Index**, the Character and Character Pattern categories include extremely useful functions for manipulating strings. This section covers only a few of the rich set of Character functions. By no means does this section completely cover the very extensive list of character functions that are available. A few things to keep in mind when you work with JMP character functions are that they typically work with both strings and lists, and, where it makes sense, it is generally possible to search in the reverse direction (end of string to beginning). Also, there are almost always multiple ways of doing the same string manipulation, using different character functions or combinations of them. Some are easier to interpret than others, so try to keep it simple.

The script for the first part of this section is 9_ParsingStrings.jsl, which includes more examples than are listed here. With the **Contains( )** function, it is possible to search a string or a list to determine whether the item of interest is within it. The case of the characters is distinguished, and if a list is being searched, the type of the item stored in the list matters. For example, in the following code the list variable tmp_lst, has items "abc", "ABC", and abc which are all different. The first two are different because of the case, and "abc" and abc are different because "abc" is a string and abc is a variable. If variables are in a list, the **Contains( )** function throws an error if the variable is not defined and returns a zero (see a3 below). This means that it is not found. The values of a variable can be searched by using **Eval List( )** inside a **Contains( )** function.

Contains

```
tmp_lst = {"abc", "ABC", 1, 3, "a", "b", [1], "abc", abc};
a1 = Contains( tmp_lst, "abc" );  //returns a1 = 1;
a2 = Contains( tmp_lst, "abc", -1 );  //Search from end to beginning
a3 = Contains( tmp_lst, abc );  //throws an error if abc is not
defined
abc = "99";
a3 = Contains( tmp_lst, abc );  //returns a zero
a4 = Contains( Eval List( tmp_lst ), abc );  // returns a4 = 9;
```

Concat Items and Concat

The **Concat Items( )** and **Concat( )** functions join strings together. **Concat Items( )** takes a list or set of strings and joins them by a specified delimiter. If no delimiter is specified, a space is used. **Concat( )** or double vertical bars (||) joins two strings together.

```
str1 = "one";  str2 = "two";  str3 = "three";
comb = Concat Items( {str1, str2, str3} );  //comb = "one two three";
comb = Concat Items({str1,str2,str3}," : "); //comb = "one : two : three";
del = ",";
comb = Concat Items( {str1, str2, str3}, del );  //comb = "one,two,three";
del = ", ";
comb = str1||Concat Items( {str2, str3}, del );  //comb = "onetwo, three";
```

Word, Item, Words

Word( ) and **Item( )** functions are similar in that they return the nth word, where a word is defined by the delimiters that are specified. A space is the default delimiter, and an empty string delimiter treats each character as a word. The difference between the two functions is that **Word( )** treats sequential delimiters as a single delimiter, and **Item( )** treats each delimiter as a separate word. The following example, which uses a quote from *Our Mutual Friend*[1] by Charles Dickens, shows the syntax of the functions. The **Word( )** function sees the comma and space (,) as a single delimiter, so it returns the second word "Charles." The **Item( )** function sees the comma and space as two delimiters, so it returns an empty string. The **Words( )** function returns a list of words where a word is defined by the delimiter.

```
/* From Charles Dickens' "Our Mutual Friend" (1864-1865) Bk.III, Ch. 2*/
Dickens_str = "Have a heart that never hardens, and a temper that
never tires, and a touch that never hurts.";

Word(3, Dickens_str);  //returns "heart", the 3rd word
Word(3, Dickens_str, "");  //returns "v", the 3rd character
Word(3, Dickens_str, "e");  //returns "art that n", the 3rd set of letters
wStr = Word( 2, "Dickens, Charles", ", " );  //returns "Charles"
iStr = Item( 2, "Dickens, Charles", ", " );  //returns ""
Words(Dickens_str);  //each word is an item in a list
Words(Dickens_str, "");  //each character is a separate word
Words("www.jmp.com", ".");  //returns {"www", "jmp", "com"}
```

Retrieve Stored Expressions

There are times when you need to retrieve an argument from an expression. The two primary functions to do this are **N Arg( )** and **Arg( )**. **N Arg( )** returns the number of arguments in an expression, and **Arg( )** extracts the specified argument. Consider the first example in the 9_ParsingExpressions.jsl script. In this example, the task is to obtain the minimum, maximum, and number of contours from a contour plot. This information is needed in order to make the contours uniform for multiple graphs.

```
cal_cp_script = cal_cp << get script;
nargs_cp = N Arg( cal_cp_script );
found = 0;  i = 1;

While( (found == 0 & i <= nargs_cp),
If( Char( Head( Arg( cal_cp_script, i ) ) ) == "Specify Contours",
speccon_cp = Arg( cal_cp_script, i );
min_cp = Arg( Arg( speccon_cp, 1 ), 1 );
```

[1]Dickens, Charles. 1997. *Our Mutual Friend.* London: Penguin Classics.

```
max_cp = Arg( Arg( speccon_cp, 2 ), 1 );
N_cp = Arg( Arg( speccon_cp, 3 ), 1 );
found = 1, i = i + 1));
```

Pattern Matching and Regular Expressions

The previous section discusses JMP character functions and the **Arg( )** function for parsing strings and expressions. These functions are tremendously helpful and very powerful. In many cases, they are sufficient to complete the task at hand. However, there are times when these functions do not have the necessary capability to do the required task, especially when mining text or reading messy data files. If that is the case, then the JMP **Regex( )** function for regular expressions or pattern matching functions might be needed. These functions are very useful when manipulating messy text files and strings. To write complex searches, it requires a lot of practice to become proficient. There are many books and other programming languages, such as Perl, that focus on manipulating text using regular expressions. If you need to do extensive text processing, you should see the references about regular expressions.

Regular Expressions

A regular expression is a concise language to describe, match, and parse text. JMP is able to process a regular expression statement using the **Regex( )** function. It uses wildcards and metacharacters to define a pattern. For complex patterns, it can be daunting. It has been said that regular expressions are write-only, meaning that once the expression is written, it is very difficult to read. That can be true, so when your script uses a regular expression, add a comment to the code that explains what the regular expression is doing.

Consider a simple example where you want to search a text string for a particular word. The scripts for this section are in 9_PatternMatchingRegExpressions.jsl.

```
//Find words in a text string
str1 = "There is a match.";
str2 = "There is no Match";
//Regular Expressions
a = Regex( str1, "match" );  //returns "match"
b = Regex( str2, "match" );  //returns . numerical missing, no match
c = Regex( str2, "(m|M)atch" );  //returns "Match"
```

In this case, the **Regex( )** function searches the strings for the word match. When searching str1, it is found, and a is set equal to match. In the second **Regex** statement that uses str2, the word Match is capitalized. So, when searching, it does not find a match, and b is set equal to a numerical missing value. In the third **Regex** statement, the matching string is changed so that it includes either a lowercase or uppercase M, and c is set equal to Match.

The following examples show how to find numeric and formatted values within a string. When using regular expressions, metacharacters are often used to define the pattern with which to match. Table 9.2 contains a subset of some of the more frequently used regular expression metacharacters.

```
//Find formatted values
str3 = "The phone number is (888) 123-4567";
Regex( str3, "\d\d\d\d" );  //returns "4567"
Regex( str3, "\d{3}" );  //returns "888"
Regex( str3, "[^(8]\d\d\d" );  //returns " 123"
Regex( str3, "\(\d\d\d\) ?\d{3}-\d{4}" );  //returns "(888) 123-4567"
str4 = "Find the value 42 in this string.";
Regex( str4, "\d+" );  //returns "42"
str5 = "Find the value 42,000 in this string.";
Regex( str5, "\d+\,\d+" );  //returns "42,000"
```

Table 9.2 A Few Frequently Used Regular Expression Metacharacters

Metacharacters	**Description**	**Example**
.	matches a single character	d.g matches "dog", "dig" , "dug"
\d	matches a digit (0 to 9)	\d\d matches any two-digit number
\D	matches a non-digit	\D\D matches "OR" , "at", "$$"
^[8]	matches any character except 8	[^8]\d\d matches "123" but not "888"
[abc]	matches any of the character in the [ ]	d[oiu]g matches "dog", "dig", "dug"
{n}	matches the previous subexpression n times	\d{4} matches any four-digit number
a|b	matches either a or b	(m|M)any matches "many" or "Many"
+	matches the previous subexpression one or more times	\d+ matches one or more digits
?	matches the previous subexpression zero or more times	red? Matches "red" or "re"

Pattern Matching

In addition to the **Regex( )** function, there are JMP functions that are focused on pattern matching. There is an extensive list of these functions described in the JMP online Help by selecting **JSL Functions ▶ Character Pattern**. In many cases, text patterns can be matched using either **Regex( )** or the pattern functions. Consider str1 and str2 that were used in the previous example. With the function **Pat Match( )**, the results are slightly different. The **Pat Match( )** function returns a 1 if matched, and a 0 otherwise. Also, it is not case sensitive.

```
str1 = "There is a match.";
str2 = "There is no Match";
//Pattern Matching
Pat Match( str1, "match" );  //returns a 1 indicating a match
Pat Match( str1, "match", "dog" );  //replaces match with dog in str1
Pat Match( str2, "MATCH" );  //returns a 1 indicating a match
Regex Match( str2, Pat Regex( "match" ), NULL, MATCHCASE );
//empty list
```

Once strings are matched and retrieved, JSL can do the analysis and manipulate the text. The following example finds and assigns variables to patterns at the beginning and end of the string, and adds what is in between to the variable middle. Note that **>?** is used to assign text to variables. Finally, it creates a new window, and writes the text to two outline boxes.

```
//Retrieving values from strings
mystring = "It is a beautiful day!";
q1 = Pat Regex( "It..." );
q2 = Pat Regex( "day." );
q3 = q1 >? z1 + Pat Arb() >? middle + q2 >? z2;
If( Pat Match( mystring, q3 ),
New Window( "Results",
Outline Box( "We matched '" || z1 || "' and '" || z2 || "'." ),
Outline Box( "The stuff in the middle '" || middle || "'." )));
Show( mystring, stuffinmiddle, z1, z2 );
```

Some Thoughts

Regular expressions and pattern matching are excellent ways to manipulate text strings and files. They are not easy to learn, and slight changes to the patterns can result in dramatically different output. If you need to use these techniques for text manipulation, our advice is to jump in and try a lot of examples. When you get the correct results, make good comments in the code about what it is that the pattern is doing to the text. Use the many books about writing regular expressions.

Use Expressions and Text as Macros

Expressions are the words, phrases, and sentences of JSL commands. An expression is something that can be calculated. An expression must be syntactically correct. When an expression is assigned, it is called a *stored expression.*

```
exprA = Expr ( 7 );             exprB = Expr( 7 + 8 );
exprC = Expr ( aa + bb );       exprD = Expr( 7 + );
```

All but exprD are valid expressions. JMP throws an error, stating that it cannot parse (interpret) 7 +. A stored expression can be simple like the previous expressions, or it can contain multiple JMP commands joined with semicolons. Stored expressions and quoted text strings can be used like macros. In other words, commands are not evaluated until they are called, you can use placeholders for values, and the stored expression can be used many times. A simple scenario to motivate the use of a macro precedes a discussion of the functions needed to run a macro and our recommended coding practices.

Framebox Customize: Value Versus Reference

Most JMP platform commands can evaluate a command with variable references. The following script snippet from 9_StoredExpressions.jsl includes four variable references: xLo, xHi, b0, and b1. This script produces the needed graph. Now, look at bivScript in the Log window or hover over it. JMP replaces variable references with their values except for framebox customizations. Chapter 8, "Custom Displays," discusses this issue. The framebox is used for adding interactive graphics, which use global variables as customizable variables.

Unless the **Add Graphics Script** command contains values (not variable references), the added curve disappears if the variables change. A stored expression like curveExpr is one solution.

```
bling_dt = Open( "$Sample_data\Diamonds Data.jmp" );
biv = bling_dt << Bivariate( Y( :Price ), X( :Carat Weight ) );

xLo = 0.2;   xHi = 2.2;
b0 = 6.5;     b1 = 1.826;

bivr = Report( biv );
bivr[AxisBox( 2 )] << {Min( xLo ), Max( xHi ), Inc( 0.1 ),
  Minor Ticks( 1 )};
bivr[FrameBox(1)] << Add Graphics Script(
  Pen Size(3);
  Pen Color ("Red");
  Yfunction( exp( b0 + b1*x), x) );
bivScript = biv << Get Script;
```

Next, curveExpr replaces the variable references with placeholders to enhance readability (i.e., _xb0_ is a placeholder for b0). curveExpr can be used to add a curve to bivr once or many times. For example, if this Bivariate analysis needed to be performed for each diamond color classification (i.e., By (:Color)), this would be a good strategy.

```
curveExpr = Expr(bivr[FrameBox(_xk_)] << Add Graphics Script(
  Pen Size(3);
  Pen Color ("Red");
  Yfunction( exp( _xb0_ + _xb1_ * x), x)
));
```

Substitute Versus Substitute Into

If an expression will be reused numerous times, use **Substitute( )** instead of **Substitute Into( )**. The first argument of **Substitute Into( )** must be a stored expression (or list) that will be changed by this function call.

```
//---If the number of globals is not an issue, store the expression
//---it makes testing and debugging issues easier.
addit = Substitute( NameExpr( curveExpr ), Expr(_xk_), 1,
    Expr( _xb0_ ), b0 , Expr( _xb1_ ), b1 );
addit;  // executes the command
```

Use **Substitute Into( )** only when building an expression or list. Otherwise, use the techniques previously shown.

1. Create a stored expression to be used as the macro (curveExpr). This makes the code more readable.
2. Use **Substitute** to create a new expression from its first argument, which is also an expression. This first argument does not need to be a stored expression because **Substitute** does not change the first argument. However, if the first argument is a stored expression (like curveExpr), use the **Name Expr( )** function. **Substitute** evaluates its first argument. The **NameExpr( )** function, when evaluated, merely returns the expression.
3. Assign the resulting expression to a variable (for example, addit). This technique makes testing and debugging much easier. As you step through the code (running line by line), you can see the value of addit using **Show**, **Watch**, or hovering techniques.
4. Execute the command. Separating expression creation from execution provides more control when testing.

The first example in the 9_StoredExpressions.jsl script uses this recommendation. It also includes an example of achieving all four steps in one command, which does not have the advantages of traceability and readability.

Text Versus Expression Macros

The technique of using stored expressions and **Substitute** functions as a macro can also be achieved with stored text and **Eval Insert** or **Eval Insert Into** functions. For both of these functions, the first argument is a string with special characters denoting the start and end of an expression. The second optional argument is the start delimiter, and the third argument is the end delimiter. Like the following example, if there is no third argument, the start and end delimiters are the same. If there is no second argument, the default delimiter **^** is assumed.

Between the start and end delimiters, a valid expression is expected. The function **Eval Insert** evaluates the expression between each set of start and end delimiters. For example, **%Char(b0)%** is replaced with the text 6.5, and **%Char(b1)%** with 1.826. In the last script line, **Parse( )** converts the string to an expression, and **Eval( )** executes it. Stored text macros more closely resemble macros of other languages.

```
//Eval Insert <-> Substitute and Eval Insert Into <-> Substitute Into
macroStr = "\[bivr[FrameBox(%Char(k)%)] << Add Graphics Script(
Pen Size(3);  Pen Color ("Red");
Yfunction( exp( %Char(b0)% + %Char(b1)%*x), x))]\";
k= 1;
curveStr = Eval Insert( macroStr, "%" );
Show( curveStr );
Eval( Parse( curveStr ) );
```

A Few Items to Note

- Large blocks of text with embedded quotation marks are easier to read and to manage by quoting the text with **"\[...]\"** versus using **!\"** to denote each embedded quoted text string.
- The delimiter **%** was chosen over the default **^** because **^** has an algebraic meaning and might be misread.

This is a simple, yet popular use of macros to embed values (not references) for framebox customization. These same techniques can be used to create custom applications and dialog boxes.

Functions: Pass By Reference Versus Value

The previous section demonstrated the use of macros (stored expressions or text), which are evaluated only when they are called and can be used many times. Remember that functions work the same way.

This section continues the discussion of reference versus value in the context of functions. It provides direction to help you with a common programming dilemma: should I use a macro or a function?

Function Versus Expression

For compiled languages, macros (stored expressions) are often deprecated for many reasons, such as optimizing code, type checking, etc. Because JSL is an interpreted language, the pros and cons are less clear. One main benefit of a well-written function is its widespread usability. Suppose your company has its own work shift calendar that can vary by even and odd workweeks. As a function, it could be an add-in, a column function, or reused in multiple scripts and shared with coworkers.

A Few Items to Note

- F: A function clearly defines the role of its arguments. M: Arguably, a text macro has an equivalent benefit. (See macroStr in 9_StoredExpressions.jsl.)
- F: A function can return a status flag. M: With proper scripting, the macro can create a status variable.
- F: A function with **Default Local** can create numerous variables. They do not collide with external variables, and memory is released once the program is executed. M: The macro can have a **Local( )** function wrapper with the same benefit.
- F: A function can be called recursively. M: Okay, a macro has no answer to that. However, a macro typically can take less memory because it is one address, not a stack of commands.

In reality, you will develop a style that is easiest for you to use. If you borrow from existing code, you will modify the borrowed code's methods. We use both approaches. Early scripts mainly used strings and expressions. Functions are used more frequently as our coding skills mature. Simple functions can be used to extend scripting options when using functions.

Pass By Value

Function myfun assumes that its argument is a vector, and it changes the second element to 10. It assigns two times its values to B, writes to the Log window the values of A, and returns the new vector, B. The value of x, this function's argument, is unchanged. A default function call passes its arguments by value. That is, a copy of the values of x is assigned to A. All calculations occur with A and leave x intact.

```
myfun = Function( {A}, {Default Local},
   A[2] = 10;
   B = 2 * A ;
   Show( A );
   B );
x = [2, 3, 4, 5];

//===Pass by Value
Show( myfun( x ), x );
// the value of x is unchanged
```

Pass By Reference

If the same function is called using a reference to the variable, Expr(x), instead of its value, x, the behavior is different. Modifications applied to A are applied to x. A copy of x is not created.

```
//===Pass by Reference

Show( myfun( Expr(x) ), x );
// the value of x is changed
```

In other programming languages, a pass by reference occurs when a pointer to an array or variable is passed, not the array itself.

When working with custom functions and very large data sets, pass by reference can make a huge difference in performance and memory usage. Chapter 10, "Helpful Tips," demonstrates this memory savings.

Return More Than One Value

This next example is one that seems to escape quite a few scripters. If you want to return more than one value, then return an evaluated list of items.

```
//=============Return Multiple Values==========================
fun2 = Function( {a, b, c}, {Default Local},
  a += b;
  b = 2* b;
  For(i=1, i<=nrow(c), i++,
    c[i] = c[i] * Random Integer(5, 10);
  );
  Eval List({ a, b, c })
);

x = Transpose(1::10);
y = J(10,1, 5);
z = y;
outList = fun2(x,y,z);
Show( outList, x, y, z);

//========== Replace values
{x, y, z} = fun2(x,y,z);
Show( x, y, z );
```

In the first call to fun2, the output is stored in outList, and the values of x, y, and z are unchanged. In the second call to fun2, x is replaced with the results of a, y with b, and z with c. Both calls are pass by value calls, meaning that duplicate copies of x, y, and z are made. The values of the input arguments are changed by the final assignment statement, not by the function.

A simple modification to fun2 looks to see whether values are passed by value or reference. If they are passed by reference, then nothing is returned.

```
If ( Type ( NameExpr(a) ) != "Name",  Eval List({ a, b, c }) );
```

The 9_MoreFunctionFeatures.jsl script demonstrates calling fun2, passing by value (like the previous example), and passing by reference. There is no perfect rule for choosing expressions or functions. Having surveyed other programmers, there is a consensus among coding standard purists—always use functions.

Call SAS and R from JSL

Of course, JMP is a division of SAS Institute Inc., so it seems natural to access the broad and deep functionality of SAS using JSL. SAS is a popular programming language and software environment for graphics and statistical computing. A motivated JSL scripter can move JMP data into SAS, perform any manipulation or analysis available in SAS, and then move the results of the manipulation or analysis back into JMP for subsequent analyses or reporting.

R is also a popular programming language and software environment for graphics and statistical computing. Its source code is freely available, and its capabilities are extended through an extensive set of free user-contributed packages, which allow specialized statistical, graphical, and reporting techniques. A motivated JSL scripter can move JMP data into R, perform any manipulation or analysis available in R, and then move the results of the manipulation or analysis back into JMP for subsequent analyses or reporting.

Call SAS

There are many ways to connect to SAS using JSL, depending on the actual location of the SAS installation. Remote, metadata-defined, or local installations can be scripted. We have SAS installed locally on our Windows systems, and we can simply connect JMP to SAS by running the following JSL:

```
SAS Connect();
```

SAS Disconnect( ) disconnects the SAS connection.

As a matter of fact, there are so many ways to interact with SAS, it's difficult to know where to start learning.

In our opinion, the best way to begin to understand how JMP scripting can interact with SAS is to connect to SAS (select **File ▶ SAS ▶ Server Connections**). Then, run, review, and understand the SAS add-in scripts that are delivered from the **File ▶ SAS** menu. These scripts are available in the sample scripts directory. They have been written by knowledgeable professional scripters to perform analyses using some of the more advanced SAS procedures on JMP data. These scripts return the output information, including ODS output, to JMP.

Because they have been developed by professional scripters, the SAS add-in scripts have great demonstrations of many advanced scripting concepts. They especially have wonderful examples of handling error conditions gracefully (some of which have been touched on earlier in this chapter).

Figure 9.2 SAS Add-In Scripts

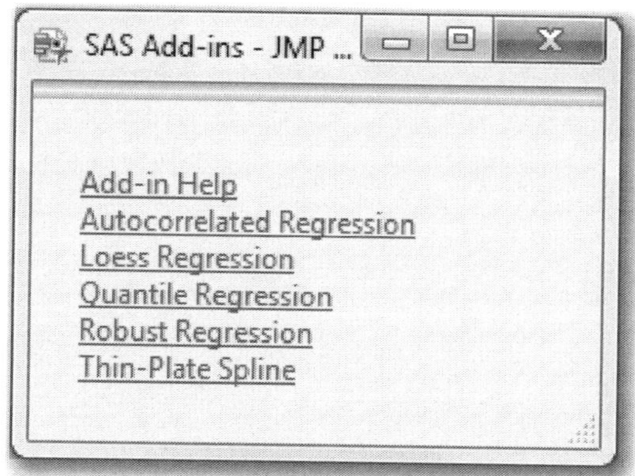

Connect to SAS, access the add-in Help file (select **File ▶ SAS ▶ SAS Add-ins**), read the procedures, run the SAS add-ins on the sample data, and admire the wonderful output that is returned to JMP. You will find many useful scripting methods that open the window to the breadth and depth of SAS!

Some words of caution: SAS has two different representations of time data. JMP has only one. JMP tries to transform time data to the applicable SAS format when moving data to SAS, which might involve just an offset or an offset and scale adjustment. This can have implications when an analysis uses a parameter estimate involving a time format from a SAS analysis in a subsequent JSL analysis. Always exercise caution to ensure that you get exactly what you expect, especially with time data.

Call R

R is not distributed with JMP. It must be downloaded from the Comprehensive R Archive Network (CRAN) at http://cran.r-project.org. R must reside on the same computer as JMP. Being very different languages, JMP tries to translate commands, data types, and the like to be understood by R. The *JMP Scripting Guide* has useful documentation of the translations required, as well as the protocols JMP uses to start and stop R. It also has some simple R code to test connections and demonstrate scripting R in JSL.

The observant reader might notice that we frequently reference the scripting of other scripters. This section is no exception. We think the best way to understand how JSL interfaces with R is to study the sample scripts written by professionals and included with your JMP installation.

One particular script to study is **JMPtoR_bootstrap.jsl** in the **Sample_Scripts** folder.

```
open("$Sample_Scripts\JMPtoR_bootstrap.jsl");
```

Earlier versions of JMP have not enabled easy bootstrapping procedures. However, the user-contributed boot package in CRAN is easy to use and powerful. The **JMPtoR_bootstrap.jsl** script contains a front end that uses column dialog boxes and matrices to generate the JMP data and prepare it for sending to R. It has a business section that initializes R, sends the data, runs the R analysis, and packages the results. And, it has a back end that gets the data from R back into JMP, and reports the results of the analysis.

Here is the commented business section of the **JMPtoR_bootstrap.jsl** script:

```
//submit to R;
R Init();              // Initializes R
R Send( bv_mat );      // Sends a matrix of data to R
R Submit(    // Submits the R code (R comments use a leading #)
Eval Insert(
"\[
    library(boot)                  # Load Boot package
# Build Function for bootstrapping statistic
    RStatFctn <- function(x,d) {
    return(^RFunc^(x[d] ^trimm^, na.rm = ^na^))}
#Call Bootstrap function
    b <- boot(bv_mat, RStatFctn, R = ^NReps^)
#Call Bootstrap CI function
    b.ci=boot.ci(b, conf = 0.95, type="basic")
   b.ciout = b.ci$basic[,4:5]
]\"
)
);  // end R Submit
b_mat = R Get( b$t );   // Gets the bootstrap sample matrix from R
b_ci = R Get( b.ciout );   // Gets the Bootstrap Confidence Intervals
```

An interesting exercise is to modify the R commands to return the BCa bootstrap confidence intervals. (The BCa confidence intervals are preferred by many statisticians.) These confidence intervals are also calculated by the R boot package. We have found the **JMPtoR_bootstrap.jsl** script to be so valuable that we have included it on our own systems as a JMP add-in. Keep in mind that there are many JMP add-ins on the JMP Community File Exchange Web site that use R packages (for example, penalized (lasso) regression, multidimensional scaling, and Chernoff faces). The JMP Community File Exchange Web site is accessed through www.jmp.com.

Call Other Programs from JSL

The previous section provided examples of extending JMP with SAS or R routines using native JMP functions to send commands and share data and results. If your operating system is Microsoft Windows, you can run other programs external to JMP using a DLL, the **Web( )** or **Open( )** functions.

A dynamic link library (DLL) is also known as a shared library. A DLL file is an executable file (.exe). A DLL file allows programs to share code and resources. For example, your company might have proprietary functions to calculate yield, cost, performance, or another indicator. By creating a custom DLL, programmers writing Excel macros, C++ programs, SAS programs, JMP programs, or R programs can load the DLL and call functions defined by the DLL.

See the JMP online Help by selecting **Help ▶ JSL Functions ▶ Utility ▶ Load DLL** for complete definitions of all messages and options. The general usage is to load the DLL, declare each exported function to be used in your script, call the function by sending a message to the DLL, and when finished, unload the DLL.

```
dllRef = Load DLL("dll_path");
dllRef << Declare Function( "fcn_name", <named_arguments> );
returnValue = dllRef << fcn_name(arg1, arg2, ...);
dllRef << Unload DLL() ;
```

The following script loads a system DLL named advapi32.dll and uses the system function GetUserNameA, which returns the user's login name.

```
nSize = 256;
myusr = Repeat( ".", nSize );
usrFname = Load DLL( "advapi32.dll" );
usrFname << DeclareFunction(
  "GetUserNameA",
  Convention( STDCALL ),
  Alias( "GetUserNameA" ),
  Arg( AnsiString, update, "lpBuffer" ),
  Arg( uInt32, update, "nSize" ),
  Returns( uInt32 )
);
rc = usrFname << GetUserNameA( myusr, nSize );
usrFname << Unload DLL();
```

The syntax for each argument is **Arg**(type, <"description">, <access_mode>, <array>). The return value description is **Returns**(type), where type is one of the following: Int8, UInt8, Int16, UInt16, Int32, UInt32, Int64, UInt64, Float, Double, AnsiString, UnicodeString, Struct, IntPtr, UIntPtr, or ObjPtr. The optional argument <access_mode> can be one of the following: input is the default value and specifies that the argument is passed by value; output specifies that the value is passed by address with the initial value undefined; and update specifies that the value is passed by reference.

For this example, two arguments are declared for the function GetUserNameA, a string and an integer. Both arguments are updated by the function call, and the function returns an integer, 1, if successful.

Search the Microsoft Web site for a complete description of advapi32.dll, an advanced API services library. After running this script, rc should be 1, myusr contains your login name, and nSize is the length of your login name.

JMP 9 has a built-in utility function to get the same information.

```
login_name = Get Environment Variable( "UserName" );
show(myusr, login_name);
```

System tasks, such as creating a directory, getting a list of files from a directory, deleting a temporary file created by a script, or checking whether a file exists, can be easily accomplished with a JMP utility function: Create Directory, Files in Directory, Delete File, and File Exists, respectively.

Prior to JMP 9, most of these tasks could only be accomplished with a system DLL. If you have inherited a script or you are reading a script from an earlier version of JMP that has numerous DLL commands, we recommend that you convert them to native JMP functions where possible. Native functions make your code more flexible and robust. Because your code is more robust, it is easier to read and debug (maintainability), and is compatible with multiple users' systems.

The 9_Web_DLL.jsl script includes another system library, secur32.dll. secur32.dll declares a function that returns the user's display name (last_name, first_name) when using option three or a unique user login ID when using option six.

There might be other programming languages besides SAS and R or more system commands that you want to run from JSL.

One method to run these other programs is to load the Microsoft Shell library shell32.dll that executes a command. For simple tasks, the JSL **Web( )** or **Open( )** commands can provide the same functionality with less code.

```
// Windows recognizes .XLS files and will open Excel for this file
web("C:\Program Files\SAS\JMP\9\Support Files English\Sample Import
Data\Demand.xls");

//This command opens a Windows command prompt
web("c:\windows\system32\cmd.exe");
```

Web( ) allows a single argument that must be a pathname. If you use the **Open( )** command and the file is of a type that JMP does not recognize, it will hand it off to the operating system to open. The trick is to specify the path of a .bat file (also known as batch file). When Windows encounters a .bat file, it executes each line of the file as if it had been entered from a DOS command prompt (cmd.exe). Commands specified in a batch file might include multiple arguments. Most, but not all programs allow command-line arguments.

To test a JSL script that calls a different type of program, you need to have the targeted program installed. The 9_Web_DLL.jsl script includes examples to zip and unzip JMP files using two executables, wzzip.exe and wzunzip.exe. These executables can be installed from the WinZip Support Web site for free.

These executables enable command-line arguments for WinZip. Here is a snippet using the JSL **Web** command. The full script documents how to use the Microsoft Shell library shell32.dll for the same tasks.

```
// Windows recognizes .XLS files and will open Excel for this file
web("C:\Program Files\SAS\JMP\9\Support Files English\Sample Import
Data\Demand.xls");

//This command opens a Windows command prompt
web("c:\windows\system32\cmd.exe");

zip_text = "wzzip c:\temp\demojsl.zip "
   || "\["C:\Program Files\SAS\JMP\9\Support Files English\]\"
   || "\[Sample Scripts\demo*.jsl"]\";
 //paths with spaces must be enclosed in double quotes for DOS

SaveTextFile("c:\temp\zipit.bat", zip_text);
wait(1);
web("c:\temp\zipit.bat");
```

If you do not want to download these WinZip commands, see the 9_Web_DLL.jsl script for the general syntax, and try the same techniques on your favorite other program.

New to JMP 9 is a Zip Archive object with native JSL functions to read, write, and get a directory listing. However, the contents of the Zip Archive object are restricted to text strings or blobs. Revisit the 2_Extra_ZipFiles.jsl script that includes an example that converts a JMP file to a blob and saves it in a Zip Archive object. Then, restore the file. See the *JMP Scripting Guide* for details and examples.

10 Helpful Tips

Introduction

Throughout this book, we include suggestions to make scripting easier and more efficient. This chapter captures those suggestions and adds a few more. Obviously, our methods and opinions are rooted in our use and JSL consulting experience. And, like most people, if we have a proven solution (a method that works with reasonable performance), we do not always look for better methods. We look forward to getting your suggestions and learning new methods from you.

This final chapter includes tips for planning your code, a watch list of common mistakes, methods that we use to find script errors (debug code), and suggestions to improve and test script performance.

Lay Out Your Code

We get JSL requests that range from simple to complex. For example, we get simple syntax questions, requests for help with debugging a script, requests for help with extending an inherited script, and requests to write a script from scratch for a specific task. If the request is more than a simple syntax question, we start asking questions. Table 10.1 documents our triage-like questions in the form of a checklist. Our intention is to help you plan your project and the layout of your script. Software developers call this task scoping or gathering requirements.

Table 10.1 JSL Project Checklist

	Data Analysis Checklist
1.0	**Usage**
1.1	Interactive user interface (dialog boxes, options)
1.2	Inactive user interface (turnkey)
1.3	Personal usage
1.4	Used by others (many, few), frequency of use
1.5	Well-defined data layout
1.6	Dynamic or conditional data layout
2.0	**Output**
2.1	Decorated JMP window or custom interactive display
2.2	Files that need to be saved, paths, etc.
2.3	Inactive graphs and reports (JMP .JRN file, Web page, PPT file)
3.0	**Data Sources**
3.1	Database query
3.2	Web source
3.3	File
3.4	Static or dynamic (query, filename, path)
4.0	**Data Quality and Data Volume**
4.1	Missing data, case, data type, modeling type, wrong values
4.2	Check date ranges, include outlier detection
4.3	How many parameters (columns)? How many cases (rows)?
5.0	**Data Preparation**
5.1	Sort, join
5.2	New columns, formulas
5.3	General statistics (n obs, start and end date, n groups)
5.4	Raw data reports or aggregated data
5.5	Stacked versus unstacked format
6.0	**Data Analysis Steps**
6.1	Partition by versus group by
6.2	Scale options (How many variables? Rows? etc.)
6.3	Advanced analyses methods, supervised (Yes? No?)

If you are new to writing scripts, you might be puzzled by items in this checklist. For example, why would the questions, "Who uses a script?" and "How often it is used?" have any influence on how a script is written? If the script requires any user input, no matter how simple it is, and even if it is run frequently, some help should be available when the user is prompted. If a script is run by the same few people on a frequent basis in a turnkey application that requires no user input, then more than likely no prompts or help are needed.

The checklist reminds us of the steps of an analysis, and helps break down a big programming task into smaller modular scripts. Suppose your coding project requires getting user input for the data query. Several data tables need to be joined, some data need cleansing, there are several analyses, and the output needs to be stored in a JMP project (*.jmpprj) file. In addition, the project requires a summary of the analyses and graphs to be reported as a Web page for a quick view to determine whether a more in-depth review of the jmpprj file is needed. Here are a few ideas to get you started.

Start with a representative set of data files.

1. Write a script to cleanse the data and join the data tables. Test it with modified versions of the data set.
2. Write the script to run the analyses and save as a JMP project. Test.
3. Write the script to save the graphs and summary for the Web page. Test.
4. Check whether the output and algorithm meet the requirements. Review with users or user champion. Refine if necessary.
5. Lay out the prompts for the queries.
6. Write the query expression as a script. Test with different hardcoded values a user might select.
7. Write the script for the user interface (UI) to get the user input. Test.
8. Integrate the smaller scripts into a parent script using **Include( )**. Check variable scoping. These options include global and local variables, namespaces, the special namespace **Here** using **Names Default to Here( )**. Test.
9. Challenge, test, and debug your script using good and messy data to ensure that the intentions are met, analyses are correct, and that it runs without issues. Ask a small core group of users to test.
10. Write training documents if needed. Create an add-in if needed.

Personally, we contend that writing robust, user-friendly UIs (especially UIs with interactive or conditional controls (if boxes, disable options, etc.)), are the most challenging and time-consuming scripts to write. By creating the UI script in a later step (step 7), after the basic report is completed, the requestor could run the script without a UI from a file while the UI script is being developed. After a review of the output (step 4), if changes are requested that require new information, the new requirements are captured before time is spent coding the UI.

The scope, the number of columns, and the number of rows in the data set influence the final script layout. For example, if the final report is a Web page or a hardcopy medium like a PDF file, with 500 or more columns and many thousands of rows, getting the data and analyzing a few columns at a time will probably save time and memory. If the final output needs to be an interactive, live chart display, create the data in an easy to access format, and create a window with a summary and controls to quickly loop through a subset of charts at a time. For all of these layouts, the code is very different.

The last section, "Performance," includes recommendations for several issues that degrade script performance.

Learn from Your Mistakes

"Always make new mistakes," which is the life motto of esteemed Esther Dyson, instructs us to not be afraid of trying and failing and to learn from our mistakes. In other words, don't make the same mistake twice. It is something to strive for, both in life and in JSL scripting. New mistakes are learning opportunities—your skills improve with each problem that you understand and solve. However, it seems that no matter how much effort you spend, your script will have repeat errors: a few typographical object misspellings and missed semicolons. The following sections present common errors that we see frequently when consulting. The list is certainly not complete and only highlights the more prevalent errors.

Format

Datetime format issues occur frequently. An entire section in Chapter 4, "Essentials: Variables, Formats, and Expressions," is devoted to dates. Recall that there are several issues that warrant this extra attention.

- Datetime values are stored and computed as the number of seconds since midnight, January 1, 1904. Recognizing that JMP is working in the background on the number of seconds since a given date, regardless of the display format, helps reduce the confusion when manipulating a datetime variable.
- Dates in JMP are numeric, continuous values. Other sources, such as Excel and databases, also use numeric representations for dates. There is no single standard for internal

representations. For example, the numerical representation for 09/25/2011 2:41:23 AM in JMP is 3399763283. In another statistical software application, its internal value is 40811.1120717593.

- Recall that JMP datetime columns have both an informat and a display format property. When saving a JMP table with datetime columns as a text file, the datetime columns are written as text strings using the display format. This is a common practice among software and database extracts.
- Reading a date string with spaces can be problematic, especially if the field delimiter is a space instead of a comma or a tab. Choose datetime formats wisely. When writing text files, we recommend using a comma or a tab for field delimiters.
- When reading a file, if JMP does not recognize the datetime format, it stores it as a string character variable. In applications that ask users to select a column for a date value, a user might incorrectly select a character column since the column values look like dates to the user.

If a column is going to have the wrong format after it is imported, it will most likely be a column with a date. Why does this matter? If a date is imported, and it comes in as character, this can cause analyses or graphs to be misleading. Consider the following script using the JMP Sample Data table Airport.jmp. This script is in 10_LearningMistakes.jsl.

```
airport = Open( "$Sample_Data\Quality Control\Airport.jmp" );
  airport_dt << Fit Y by X(
    Y( :Delay ), X( :Day ),
    X Axis Proportional( 0 ));
```

Figure 10.1 Airport Data—X Axis Is Day, a Date String

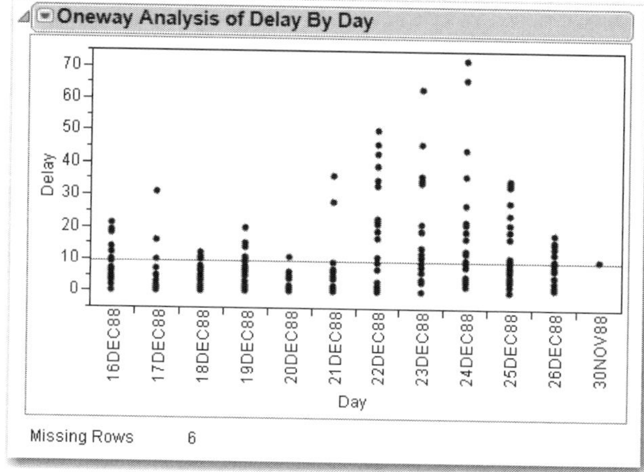

In this case, the variable Day has the format ddMONyy, which is not a format that JMP recognizes as a date. If an additional row is added to the table for data with a November date, it plots that new point at the end of the graph, in alphanumeric sequence, not in date sequence. Also, because Day is a character date, its modeling type is nominal, and the **Fit Y by X** command returns a **Oneway** Analysis. Subsequent messages to customize an expected **Bivariate** object will throw error messages to the Log window. Scripts that read dates or use them in analyses require extra attention, including type checking, formatting, and error capturing.

Another common format issue occurs when the JMP modeling or data type is incorrect. Several JMP platforms, such as **Fit Y by X**, are context-sensitive to the modeling type of the specified variable; that is, the modeling type determines the resulting output. An unintended (incorrect) modeling type creates unexpected results. Similarly, an incorrect data type produces an error for any platform with strict data type requirements. The 10_LearningMistakes.jsl script includes several type checking statements. They are **Get Data Type** for column Day, **Get Class Name** for the object returned by **Fit Y by X**, and the **Try( )** and **Throw( )** functions to capture errors. The following snippet demonstrates commands to convert the ddMONyy date format. Figure 10.2 includes an alternate view of Figure 10.1 when the X axis is a true date value.

```
//If the format is known to be ddMONyy, convert
mon_AA = ["JAN"=>1, "FEB"=>2, "MAR"=>2, "APR"=>4, "MAY"=>5, "JUN"=>6,
      "JUL"=>7, "AUG"=>8, "SEP"=>9, "OCT"=>10, "NOV"=>11, "DEC"=>12];
:Try(:Date <<Delete Formula ); //Delete if it exists
:Date << Data Type("Numeric");
:Date << Set Modeling Type("Continuous");
:Date << Set Each Value( As Date(
       Date DMY(
         Num( Substr( :Day, 1, 2 ) ),
         mon_AA[Substr( :Day, 3, 3 )],
         1900 + Num( Substr( :Day, 6, 2 ) )
       ) //end DMY
     )  //end As Date
); //end set each value
:Date << Input Format( "d/m/y" ) << Format( "m/d/y", 12 );

Try(biv = airport_dt << Bivariate( Y( :Delay ), X( :Date )),
//if errors
   Caption("Either X or Y is not numeric..aborting");
   wait(5);
   Caption(remove);
   Throw( "Non-numeric input columns")
);
// if no errors
bivr = Report(biv);
bivr[ScaleBox(1)] << {Interval( "Day" ), Inc( 1 ),
Minor Ticks( 0 ), Rotated Labels( "Angled" )};
bivr[FrameBox(1)] << {Frame Size( 431, 195 )};
```

Figure 10.2 Airport Data—X Axis Is Date, a Numeric Date

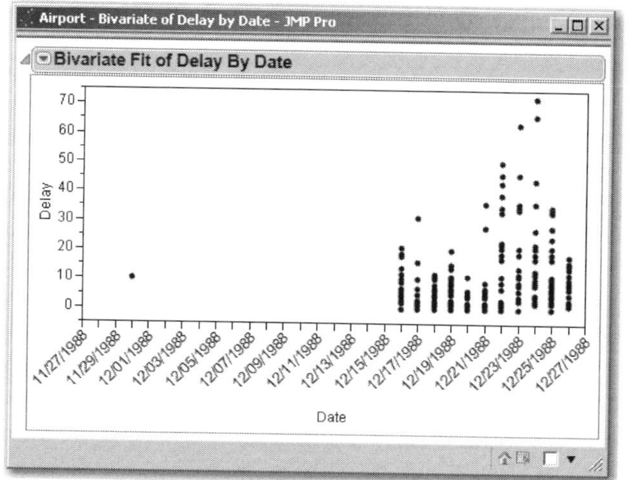

Syntax

Scripters make many syntax mistakes. It is fairly common to have these types of mistakes, and JMP usually provides helpful error messages when they are encountered. Here is a list of common syntax mistakes:

- Incorrect spelling of platforms, functions, etc.
- Unmatched parentheses, braces, brackets (set matching fences in JMP Preferences)
- Unmatched quotes (use script editor coloring in Preferences)
- Missing semicolon or dangling colon (be careful when copying scripts from data tables)
- Missing comma between options
- Extra comma after the last option

For some syntax problems, the Script Editor and available options in JMP help with resolving the problem quickly. Figure 10.3 shows some of the available options in the Script Editor. For help with debugging a script, it is highly recommended that **Show line numbers** is selected, and it is also useful to select **Auto-complete parentheses and braces**. The Script Editor has special coloring for functions, text, comments, etc. For example, if you are using the standard colors and all of your code appears magenta, JMP is interpreting your new command as a string. In other words, you have an unmatched double quotation mark. If all of your code appears green, look for an unmatched comment (**/*** with no matching ***/**). If you are typing a function and it is not blue, look for a spelling or syntax error. For example, **Files in Diretcory("$Sample_Data")** will not be the correct color until you transpose the **tc** to **ct**.

Figure 10.3 JMP Script Editor Preferences

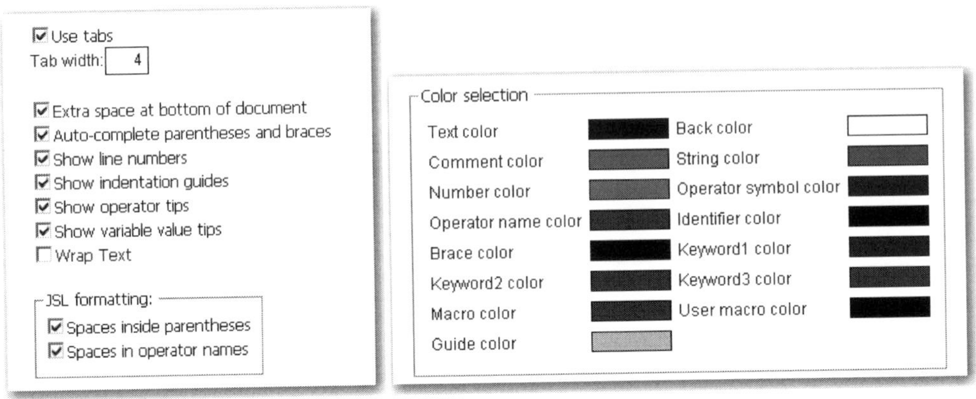

Programming

Programming mistakes can be more subtle and difficult to discover and fix. An important and often overlooked part of writing a script is the planning stage. It is important to understand the scope of the solution and the audience. A script that will be used only by the scripter on a single file type will likely be written very differently than a script that will be used by many on a variety of files and file types. Here are some common issues encountered:

- Working on the wrong data table (always be explicit about the data table that is being used).
- Using syntax from one platform and applying it to another.
- Using a column reference such as :age when the syntax requires a quoted string "age".
- Variable scoping conflicts.
- Incorrect program flow.
- Incorrect row number.
- Scriptable[ ] is written to the Log window. This can mean one of two things. There is no output from the script, and JMP is acknowledging that the script ran. If output is expected, there is an error in the code if no output is given.

Error Checking

When writing a script, you should include error checking. This adds more complexity to the script and the script takes more time to write and run, but in the end, it is worth it. It makes the script robust (or at least more robust!) to what others throw at it. Use the **Try( )** and **Throw( )** commands for error checking and to make a script more robust, but remember to be selective.

We have seen scripts where one large **Try( )** statement was applied to an entire script, and another script where a **Try( )** statement was applied to every statement. A single **Try( )** has the downside of not flagging where in the script the error occurred, which is not useful to the user or the script owner. Run the following script and check the Log window. Then, highlight the statements for A through E, and run again.

```
Try(
    A = Sqrt( 0 );
    B = Sqrt( 2 );
    C = Sqrt( -9 );
    D = Sqrt( "4" );
    E = Sqrt( 4 );
,   //else
Throw( ) );
```

Too many **Try( )** statements can create a false sense of error capturing or might reduce the script's performance. One scripter thought the statement Try(y = sqrt(x), Throw()) would catch an error if x was less than 0. For more details, see examples 5 and 6 in the 10_LearningMistakes.jsl script.

Debug Your Scripts

Wouldn't it be thrilling to write a large and complex JSL script and have it run exactly as intended the first time? Based on some of our experiences, this happens rarely enough that it, indeed, would be remarkable. Some of us have problems writing multiple lines of scripts without an unintended consequence rearing its head. *Debugging* means finding, diagnosing, and fixing errors in your script so that your intentions are realized in the output. As industrial statisticians, we enjoy many opportunities to debug not only our own scripts, but those of engineers with whom we consult. As a result, we are sharing some of the systematic methods that we have developed and used in our own work.

Your best friends for finding and fixing problems are the Script Window and the Log window. In the Script Window, you can hover over a variable to get values and to ensure that your script is functioning as intended. We find line numbers and colors to be very useful.

Some scripts fail to even interpret, and JMP throws a JMP Alert window. This window typically has a helpful suggestion about where to look and what to look for.

Figure 10.4 JMP Alert Window

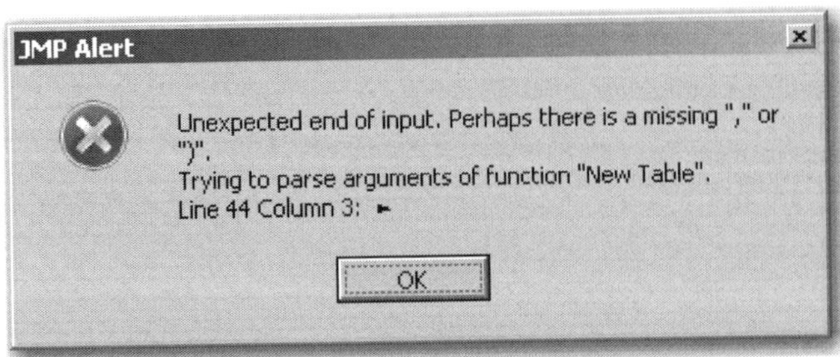

In the Log window, one of the more obvious signs of a problem to fix is when you get pounded. Syntax errors are often highlighted with three sequential commented pound signs. When the script is run that produces the error in Figure 10.4, the error is marked with **/*###*/**. These pound signs are inserted at the location where JMP believes it has encountered the error. Usually, we find that getting pounded results from typographical errors or misplaced bracketing. If your Log window is very long, searching for the pound signs from the **Edit** menu can work efficiently.

In our experience, boldly submitting large scripts without incremental testing is rarely rewarded with timely success. We tend to sequentially highlight small portions of a script, run it, and then fix any errors of commission or omission that we might have introduced. When we are satisfied that our results are correct in the small portions, we clear the Log window and run the entire script. Next, we validate the results of the entire script. The ability to run small blocks of code is one big advantage of JSL. We strongly recommend the use of this feature when debugging your scripts.

The **Wait( )** command enables you to pause script execution for a value of time, and then continue running the script. With some system tasks, especially accessing files over networks or publishing Web pages to separate servers, a **Wait( )** command might be necessary to successfully fulfill your intentions.

We do not recommend submitting complex **For( )** loops with a large number of iterations without incremental testing. When scripting **For( )** loops, we tend to ensure that the script fulfills intentions with a small number of iterations, such as i<=2, and then we expand the iterations. Sometimes, we highlight i=1 and run each command (up to the semicolon) to find the culprit.

Include( ) commands are very useful, especially for modularizing more complex scripts. However, debugging might be more difficult when using included scripts, especially unfamiliar scripts with many functions. We have found that opening the included scripts and performing our same incremental approach to debugging can save significant time and avoid frustration.

It is a wonderful scripting practice to throw informative error messages when an **Include( )** file encounters a problem. When many **Include( )** commands are used in a single parent script, you can save many hours of troubleshooting by embedding information in the error message, such as the Include filename or a reference to which file erred where.

If you encounter a single line of script that is problematic and confusing because of its length or complexity, you can sometimes break that single line into two or more lines for easier debugging.

The Windows instantiation of JMP v9.0.2 includes a debugger that you can invoke by inserting the command /*debug step*/ in the first line of your script. Unfortunately, this command in the first line of our scripts caused irretrievable exceptions that made JMP terminate prematurely, so we cannot evaluate or recommend the debugger. (Although, from that experience, we can recommend saving your scripts regularly.) We have been told that this error occurs only with certain background themes in Windows 7, and that future enhancements to the debugger are planned in upcoming versions of JMP. A helpful debugger will prove very useful and is eagerly anticipated!

Two of the most egregious errors that we encounter are an invalid row number error and the Scriptable[ ] error. Finding and fixing them can be vexing.

An invalid row number usually occurs when assignments are made to ambiguous names, and they manifest themselves in unintended missing values. It has proven difficult to find the sources of naming ambiguity in more complex scripts. Using strong names and proper scoping reduces naming errors. For debugging this error, creating a list of column names and global variables will help pinpoint the sources of ambiguity.

While debugging this error, it is important to understand that the current row setting lasts only for the portion of a script that you select and submit. After a script executes, the current row setting resets to Row 0. This behavior can make a submitted script all at once produce different results from the same script submitted a portion at a time. So, even if a script runs flawlessly when submitted in small portions, be sure to run the entire script and ensure that your intentions are fulfilled.

The Scriptable[ ] message in the Log window might not mean anything bad if you don't expect any output. However, if you anticipate output, and you get the Scriptable[ ] message in the Log window, it means you have written a syntactically correct script whose programming logic has gone horribly wrong. You will receive no hints (not a single clue, whatsoever) about where or how. Please note that this lack of information is the fault of the scripter, not JSL. After all, the script syntax is correct, but JSL cannot interpret your intentions. For that reason and for how incredibly frustrating it can be to fix, the Scriptable[ ] error is one of our least favorites. Our experience with debugging Scriptable[ ] errors suggests that if you can sequentially narrow down the problem to one section of your script, you are more efficiently served by going back to the drawing board. Sit down and carefully plan and rewrite that section of your script, while concentrating on the logical ordering and implementation of the scripted operations.

Do not be afraid to ask for help. JMP Technical Support is a consistently excellent resource, and there are many JMP groups, blogs, and discussion sites on the Internet.

Performance

For small or reasonably sized data files, the performance of a script is usually not an issue. In these situations, getting the script to give you the appropriate output can be the only concern. As technologies improve and computing power increases, it is easier to collect data. The amount of data collected is growing rapidly, and data table sizes are increasing. For large files or complex manipulation, the performance of a script is important. There are a few key things to remember that will improve the performance of your script. JMP provides commands for memory usage and timing to help track performance.

Some Tips

Creating new tables consumes resources, so reduce the number of tables you create when possible. To avoid generating new tables, use the following strategies:

- Create matrices and lists rather than new tables.
- Use the **Summarize** function rather than the Summary platform.
- Use the **Invisible** option for tables (or reports). Be sure to close tables and reports when they are no longer needed.

Input and output are not free, so, when possible, limit it. This is related to table creation, but extends to other areas as well.

- When debugging a script, it is common to use **Show( )** to reveal variable names. If there is a **Show( )** function in an iterative loop, and the show information is being written to the Log window at each iteration, this can seriously increase the time that it takes the script to execute. Once the script is written, remove or comment out the show statements. For large scripts, use a conditional show statement to save time. See example 1 in the file 10_DebugYourScripts.jsl.
- Any change to the data table forces the display to update. Use **Begin Data Update( )** and **End Data Update( )** to control the table update. This allows blocks of updates to be done

once. It is important that **End Data Update( )** is done or the display is not updated until it is forced.

- Where possible, create graphs only once. Avoid having to redraw them. When there are many points, graphs take time to draw.
- Avoid using column formulas when possible. Often only the formula values are required, not the formulas. Unless you really need the formulas, use **Set Each Value( )** instead. If a formula is required, control its evaluation by using **Suppress Formula Eval( ).**

When reading files from or saving files to a network drive, allow time for the completion of the task by using **Wait( )**. This slows the script, but if the remainder of the script requires complete files, it is a necessary delay.

Test Performance

There are times when you might want to measure the performance to help optimize a script. The two most common performance statistics for software are memory and time. Two functions in JSL to test memory usage and timing are **Get Memory Usage( )** and **Tick Seconds( )**.

When testing memory usage, it is important to understand that JMP controls memory allocation, and memory usage depends on the system and what is currently running. For example, defining ten new variables might not increase memory usage if previously allocated memory is available. However, it is safe to say that the memory usage is probably different after opening or closing a large file or creating a matrix with one million rows.

The 10_MemoryPerformance.jsl script provides different scenarios for testing memory usage, such as the increase in memory when a large table is opened. Here is a snippet of the script that demonstrates how to obtain the memory usage before a large table is created, after creating it, and after closing it. The differences between before and after table creation, and before and after table open and close, are written to the Log window.

```
Clear Globals();  Delete Symbols();
m0 = Get Memory Usage();
dt = New Table( "Tall",
New Column( "X1", numeric ),
New Column( "X2", Numeric ),
New Column( "X3", numeric ),
New Column( "X4", Numeric ),
New Column( "X5", numeric ), Add Rows( 2000000 ));
:X1 << Set Each Value( Random Integer( 1000, 1999 ) );
:X2 << Set Each Value( Random Integer( 2000, 2999 ) );
:X3 << Set Each Value( Random Integer( 3000, 3999 ) );
:X4 << Set Each Value( Random Integer( 4000, 4999 ) );
:X5 << Set Each Value( Random Integer( 5000, 5999 ) );
dt << New Column( "Param", Character );
```

```
dt:Param << set each value( "P"||Char( Floor( (Row() - 1) / 1000 ) +
1 ));
Wait( 0 );
m1 = Get Memory Usage();
Close( Data Table( "Tall" ), NoSave );
m2 = Get Memory Usage();
Show( m1 - m0, m2 - m1 );
```

The scripter has some control over the time it takes a script to execute. For large data files, it is important to write code as efficiently as possible. Using the **Tick Seconds( )** command can help with testing the efficiency of different algorithms. The script 10_ReadWritePerformance.jsl provides different scenarios for testing the time it takes for some simple tasks. The 10_AlgoPerformance.jsl script looks at testing different algorithms for a simple column calculation and demonstrates that the timing differences can be dramatic. Consider testing the accuracy of the **Wait( )** function.

```
Clear Globals();  Delete Symbols();
t1 = Tick Seconds();
Wait( 10 );
t2 = Tick Seconds();
Write( "Accuracy of Wait 10 ", Round( t2 - t1, 3 ), "\!r" );
```

Pass By Reference Versus Pass By Value in Functions

When using functions, arguments can be passed by value (the default) or by reference. When passing a variable by value, a temporary copy is made in memory taking up space and making changes to the local variable. In other words, the changes are not visible outside of the function. When passing a variable by reference, the function operates on that variable and all changes are visible once the function is complete. This can be critical when working with large data files in complex scripts. The 10_PerformancePass_Ref_or_Value.jsl script provides a simple example. To pass an argument by reference, simply treat the argument as an expression, and enclose it in the function **Expr( )**. For example, in the following two lines of code, the first line calls the function **foo** using references to variables, and the second line calls it using the values of the variables:

```
//passing arguments by Reference using Expr
foo( Expr( bigx ), Expr( bigy ), Expr( bigz ), N Col( bigz ) );

//passing arguments by Value
foo( bigx, bigy, bigz, N Col( bigz ) );
```

Check the Log window after running each block of code. After block 1, which uses pass by reference, the memory usage is about the same as before running foo, while foo is running and after foo returns control. After block 2, which uses pass by value, the memory usage temporarily increases while foo is running. This increase in memory, even though temporary, may exceed your computer's capacity and cause your program to abort.

A Few Items to Note

- Algorithm efficiency is the average time necessary for an algorithm to complete. It is often characterized by its order, called big O notation. For example, a double nested For loop is defined as $O(n^2)$. A double nested For loop is an acceptable method for reasonably sized data sets. For very large data sets, that same nested For loop might run out of memory. Or, an $O(n^2)$ algorithm that takes 0.001 sec to complete for n=1000 may take 1000 seconds for n=1,000,000. There are many algorithms for handling very large data sets. Our goal is to make you aware of this issue, we recommend that you get and set program requirements and then enlist the aid of a computer science expert with algorithm and scaling experience.
- If you are new to scripting and your data set is not very large, keep it simple. Learn to use functions before worrying about pass by reference or pass by value or algorithm efficiency. If your data sizes are very large, get help.

Index

A

B

C

D

E

O

P

T

Y

Z

Symbols

5601270R00176

Made in the USA
San Bernardino, CA
13 November 2013